Float Your Boat!

Float Your Boat!

The Evolution and Science of Sailing

MARK DENNY

The Johns Hopkins University Press
BALTIMORE

Printed in the United States of America on acid-free paper
9 8 7 6 5 4 3 2 1

The Johns Hopkins University Press
2715 North Charles Street
Baltimore, Maryland 21218-4363
www.press.jhu.edu

Library of Congress Cataloging-in-Publication Data

Denny, Mark, 1953–
Float your boat! : the evolution and science of sailing / Mark Denny.
p. cm.
Includes bibliographical references and index.
ISBN-13: 978-0-8018-9009-3 (hbk. : alk. paper)
ISBN-10: 0-8018-9009-8 (hbk. : alk. paper)
1. Sailing. 2. Sailboats—Design and construction. 3. Sails—Aerodynamics. I. Title.
VK543.D45 2008
623.88—dc22 2008013283

A catalog record for this book is available from the British Library.

To my sisters, Fi and Sal,
and to those they love: *C*, *x*, and *w*

Contents

Acknowledgments

I am grateful to Fiona and Clayton Lewis for taking time out from sailing the Caribbean (it's tough, but somebody has to do it) to read an early version of the manuscript. Thank you both for providing useful suggestions and essential corrections. Any remaining errors are mine alone.

For the many unusual pictures that were needed for this enterprise, I have to thank a number of sources from Australia, the British Virgin Islands, Bulgaria, Canada, Hong Kong, Italy, the Netherlands, Russia, the United Kingdom, and the United States—it seems that the interest in sailing ships is worldwide. I thank the following institutions for contributing images: the Art of the Age of Sail (John Andela), CoastalBC.com (Cam), the Columbus Foundation BVI (A. J. Sanger), the Cutty Sark Trust (Anna Somerset, to whom thanks for pointing out a possible family connection that I might have with the Cutty Sark construction), DHD Multimedia Gallery (Bruno Girin, Damon Hart-Davis), the Leo Baeck Institute (Miriam Intrator), NASA (http://visibleearth.nasa.gov/), the United States Navy, Wikipedia, www.gore3d.com (Antonis Kotzias), and www.shipmodels.ru (Slava Petrov). Several individuals have generously contributed their own private photographs: for these I thank Stefania Bocheva; Mike Cawood; Jan van der Crabben; Darillo; Clayton and Fiona Lewis; Simona Manca (grazie, Simonetta, per le fotografie meravigliose); Sanjay Pindiyath; Steve Priske; and Rich Swanner. Carolyn Moser copyedited the text heroically—many thanks.

Float Your Boat!

Introduction

There is enormous interest—and participation—in sailing. Those nations with a strong maritime history have, for the past century or two, continued the ancient skills of building and operating sailing vessels. Though sailing ships have lost their place as engines of trade and weapons of war, they continue to hold a fascination for millions of people, spawning a latter-day revival of sailing technology and development. Much of this modern hi-tech impetus comes from the drive to win prestigious and glamorous international yacht competitions such as the America's Cup, where sleek and very expensive racing machines accrue all the glamour and kudos given to the nineteenth-century clippers, and for the same reasons: they are beautiful machines, and they win money and status with their speed.* Most of the interest in sailing, however, comes from weekend captains, who get away from the office to enjoy a few hours in the company of *Lady Susan* or *Gloria II*. These ladies may not be as sleek and elegant, or as fast and expensive, as America's Cup winners such as *Alinghi* and *America3*, but they are just as pampered and just as loved.

In the United States alone an estimated 4.1 million people participate in recreational sailing. Some of these sailing buffs are scientists and boat

*Yachting has figured prominently in the Olympic Games since their inception in modern form (1896). By the 2000 Summer Olympics in Sydney, Australia, there were nine classes of sailing competition.

designers, and some of them have written books about their hobby. Any physicist who sails is immediately prompted by Mother Nature to ask himself or herself how sailboats work. Many nonscientists who sail have similar questions, and they (very reasonably) ask for answers that are not couched in the mathematical jargon of physicists.

The large and growing library of sailing literature is divided into two broad categories: books that appear to be written by engineers for America's Cup boat-builders, and books that have been written by experienced boat-handlers who want to pass on their experience to weekend sailors. Books in the first category will tell you about the strength-to-weight ratio of synthetic running-rigging lines or the optimum sail angle-of-attack to avoid separation bubbles. The emphasis is on sailing as fast as possible, reflecting an increasing emphasis on yacht racing. Books in the second category will tell you about dinghy maintenance, boat electrics, the fundamentals of boat handling, and nautical etiquette—emphasizing the practicalities that must be mastered by every recreational sailor.

There seems to be something of a gap in the literature, which this book aims to fill. Books in both of the categories that I have outlined (parodied, some would say, because I am exaggerating the types to make my point) pay only nodding attention to the evolution of sailing craft and the technological or scientific reasons for this evolution. This history is interesting, and is a significant part of the story that I tell in the pages to follow. History sets our modern sailing vessels in context: there are imperative historical and technological reasons why large sailing ships and small sailing boats evolved in the way that they did.

More important, the level of technical presentation of sailing physics in the two categories is, I find, somewhat unsatisfactory. There are glorious exceptions, but generally the books that seek to explain sailing physics are either too technical—drowning readers under oceans of statistics concerning materials science, or dazzling them with advanced computer simulations—or too simple. The first case is unsatisfactory because too often the average reader will get left behind in the wake, perhaps marveling at the erudition but without having been much enlightened. The second case is often more enlightening but too often also rather misleading. A number of misconceptions about the physics that underpins sailing have found their way into the popular literature—particularly concerning the reasons for aerodynamic and hydrodynamic lift.

What is needed, it seems to me, is an *explanation* of the physics that underlies sailing phenomena: not large and complex computer calculations that reproduce every observed detail but shed little light, and not airy analogies that misinform. In the pages that follow you will see how the principles of physics are applied to explain how wind drives old square-rigged ships and modern Bermuda sloops, how torque determines stability, why hull speed exists, how sailboards and iceboats go so fast. And why lift occurs. I will sacrifice perfect accuracy on the altar of understanding, meaning that I will provide you with the single guiding principle behind a given topic (wave drag determining hull speed, for example) and show how this principle explains the phenomenon. In the real world there may be other contributory reasons for the phenomenon, but they merely cloud the issue; the basic principle accounts for most if not all of what we observe.

I will not water down the physics but will belay the mathematics.* Quite a few equations will appear in this book, without detailed derivation. If math, to you, is like close-hauling in a heavy sea, then bear away: the text alone will provide a coherent picture. If, on the other hand, you like math, then there are enough equations and explanations here (mostly tucked away in notes at the back of the book; the footnotes are reserved for more general comments) for you to reconstruct the detailed derivations. I am aiming for simple physical explanations combined with readability and will settle for approximate answers. Fine detail is missing, particularly for more technical topics such as vortex formation and shedding. If you crave the fine detail, you will need to arm yourself with some heavy mathematical tools and wade into the considerable technical literature.

The difficult question of aerodynamic lift is addressed in stages throughout the book, and my approximate approach is quite sufficient to provide a broad-brush explanation of how wind provides the motive power for sailing vessels. However, given the disparity of opinions concerning lift, and the sometimes rancorous disagreements that pepper the sailing magazines on this thorny subject, it seems like a good idea for me to devote an appendix to the subject. Here you will see what the problem is and how the misconceptions arose. Popular explanations of lift force are like the three blind men who try to picture an elephant by touching

*Mother Nature speaks the language of mathematics, but most people don't. A significant aspect of my task in writing this book is to act as interpreter.

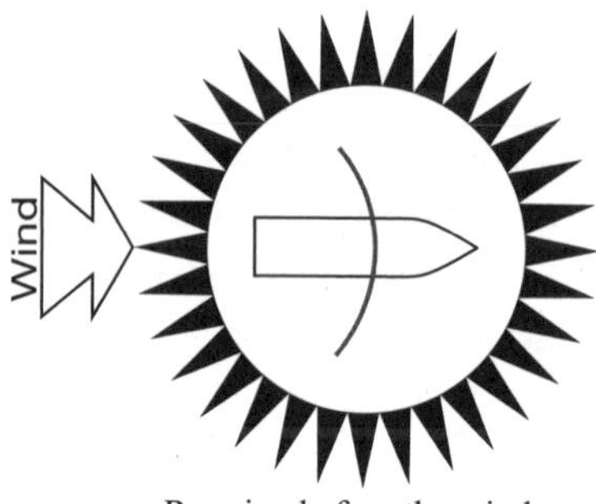

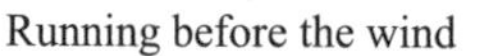

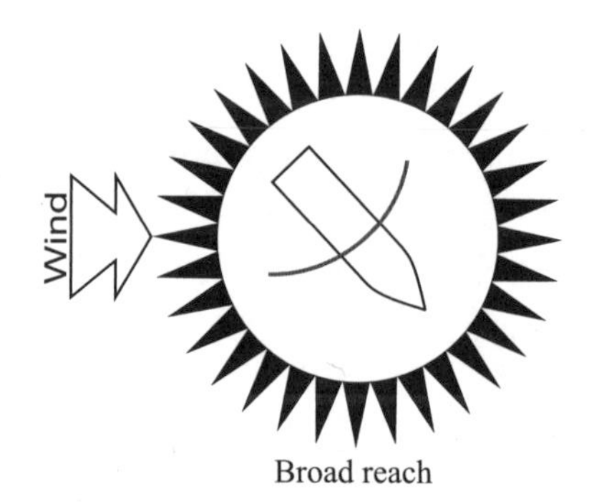

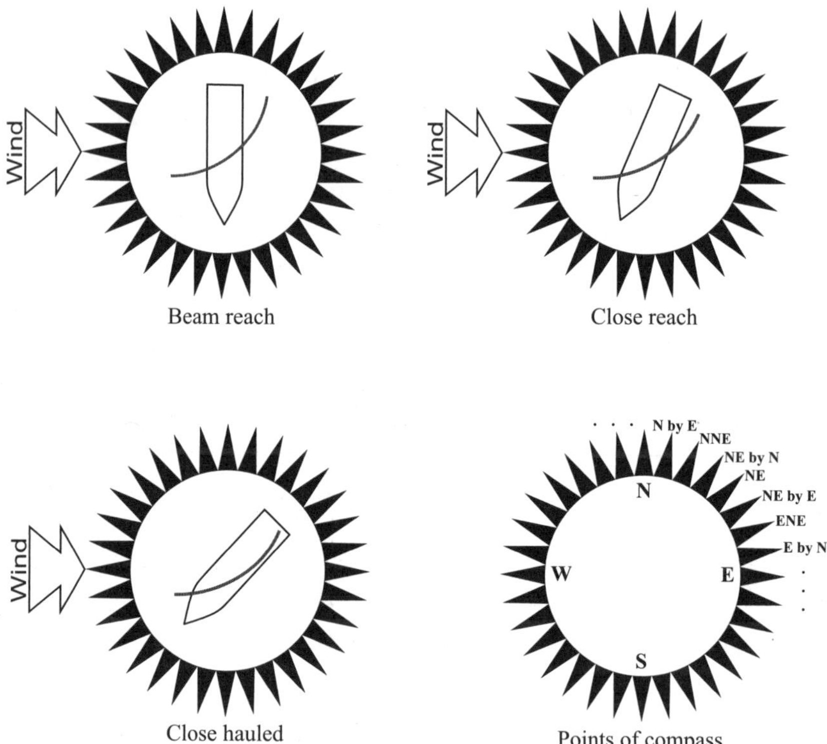

Figure I.1. Sailors give different names to the direction, relative to the wind, in which a sailboat is heading. They also divide the compass into 32 points, as shown.

different parts of the beast. Each part gives an incomplete view that cannot be extrapolated to the whole picture. An elephant is not simply an extended trunk or tail, and aerodynamic lift is not simply an application of Bernoulli's equation.* The real reasons for lift are well understood by physicists, and have been well understood for a century, but because the physics is complicated, it has proved difficult until quite recently to conjure up an intuitive explanation for non-experts without throwing overboard the essentials.

To summarize: This book is about the physics of sailing vessels. I will share with you the history of sail, paying particular attention to the technological limitations and breakthroughs that shaped this evolution. I will provide explanations for the physics behind a number of sailing phenomena. These explanations get to the core of the matter, without frills. Math is used: you can read around it without missing the plot, or delve into the endnotes to see where it comes from.

I have provided a glossary to define nautical terms and terms from physics. To start you off with nautical terms, figure I.1 shows you the *points* of sail.

Throughout the book at the beginnings and ends of chapters I've provided capsule glossaries of common words and phrases that have nautical origins but now have quite different meanings—they have lost their moorings, so to speak. These attest to the influence that sailing has had on the English language and hence on us all, even the most determined landlubber. Thus:

The whole nine yards: *Everything.* Mainsails on square-riggers were supported by yards (horizontal spars). If all three mainsails on a three-master were unfurled, then every mainsail was deployed.

You may be surprised at how many common phrases and expressions have a nautical origin. Perhaps, given the spread of the language from its island origin, this is not so surprising. As the old proverb goes, "There is plenty of sea in the Englishman's blood, and Englishmen's blood in the sea."

*Indeed, as we will see, one modern expert thinks that lift should first be taught to students without ever mentioning Bernoulli's equation.

Finally, you will find works of authors cited in the text in the bibliography at the end of this volume.

Whether you are an inquisitive landlubber who has never set foot in a boat, a student of life who likes to potter on the water at weekends, or an old salt who lives for the sea, I hope that by the end of this book you will be able to say, "So *that's* how it works!"

1

Evolution: From Prehistory to the Age of Sail

Afloat: *Floating on water.* Given the maritime tradition of Vikings and their influence on English history, the English language has a lot of nautical words borrowed or derived from these energetic Scandinavians. *Flota* is Old Norse for water. (Other derived words are *ahoy, lake, ship, keel, skip,* and *ferry.*)

Bigwig: *Important person.* In the seventeenth and eighteenth centuries, senior officers in the Royal Navy wore prominent wigs; such officers were disparagingly referred to as "bigwigs."

Coasting: *Move by momentum, effortlessly.* A coasting ship sailed close to the shore; this practice was easier and less dangerous than open ocean sailing.

Wallop. *Hit hard, thrash.* In sixteenth-century England, King Henry VIII supposedly ordered Admiral Wallop to devastate the French coast, following a French raid on the English town of Brighton.

Three Maritime Traditions

In the course of history at least three major and more or less independent traditions of boat-building and sailing have emerged. I will call these the South Pacific, the East Asian, and the Atlantic traditions. My goal in this book is to share with you what I have learned over the years about modern sailing vessels. In this chapter, I concentrate overwhelmingly on the Atlantic tradition because most modern yachts (with the notable exception of catamarans) evolved from older craft that originated along the coastal regions of Europe. Although much of the early European contribution to sailing vessels happened in the Mediterranean Sea, I refer to the evolution of European sailing as "Atlantic" because that is the area from which European ships emerged to spread around the world during the fast-changing Age of Exploration. This key period (around the year 1500 CE) saw the European "discovery" of North America and the European circumnavigation of the globe.

It would be churlish of me, however, to ignore the other sailing vessel traditions. I want to avoid the mistake of a book in my library entitled *The Oxford History of the Classical World,* published by Oxford University Press. Now, it is certainly true that Oxford University has seen many preeminent scholars pass through its cloistered quadrangles, as well as many more ordinary mortals such as myself—and no doubt a number of rascals as well. Whatever the hue of its members, one common criticism of this venerable institution is that it exudes a sense of superiority and a sort of inward-looking arrogance. I don't want to overstate the point, because this environment is rapidly changing these days, but my OUP history book epitomizes the self-centered old-school attitude: the "classical world" of the title means ancient Greece and Rome, and nowhere else. ("Classical Antiquity" has become the common expression—not just at Oxford but throughout the English-speaking world—for ancient Greece and Rome.) The unstated assertion is that classical history (roughly the millennium from 500 BCE to 500 CE) amounts to the history of these two—admittedly important—civilizations, and no others. Forget about China, India, Africa, and the Americas of this period; before Europeans got there, the rest of the world didn't have any history.

I do not wish to be accused of such a blinkered point of view, and yet I need to focus on the origins of modern sailing vessels, which are rooted in Europe. So, after an excursion into prehistory I devote a couple of sections to catamarans of the South Pacific tradition and junks of the East Asian tradition. These essays will of necessity be cursory: I am not writing an encyclopedia. It seems to me, however, that I should at least place the European nautical achievements in context, so please regard my discourses on the evolution and characteristics of catamaran and junk as appetizers which I hope will tempt you to indulge in a main course on these tasty topics at a later date.

The Prehistory of Waterborne Travel

The earliest forays of humankind onto the waters of the world occurred perhaps as long ago as 16,000 BCE in Europe, Asia, and Africa. These tentative excursions launched watercraft of four distinct types.

Dugouts. Hollowed, shaped logs formed strong boats, of a size that was determined by the available trees. Dugouts were widespread throughout the prehistoric world, and it is thought they gave rise to most modern boats as follows. Extra logs or planking would be added to the sides of the dugout to provide extra freeboard.* This is the basis for boats that are built up from a strong central keel. Dugouts almost certainly represent the background to the Atlantic tradition of boatbuilding. The oldest known dugout, found in the Netherlands, dates from 6000 BCE.

Rafts. A raft is a watercraft that floats by virtue of the constituent parts being less dense than water: rafts do not make use of Archimedes' principle.[1] Rafts are made from logs, skin bladders, or reed bundles. They are not so robust as dugouts, and yet they can venture into calm seas. It is likely that the ancient Egyptian boats and ships emerged from a tradition of raft-building because Egyptian boats did not have a keel. The Polynesian catamaran is technically a raft, and it most certainly was a very capable ocean-going craft. It may be that the Chinese junk can count the raft as its distant ancestor because junks also do not possess structural keels.

Skin boats. Skin boats consist of animal hides stretched over a skeletal frame of springy wood or bone. There are archaeological remains of vessels made from reindeer skin and antlers dating back to 9000 BCE. More modern examples include the Inuit kayak, the Irish curragh, and the Welsh coracle. Such vessels are relatively frail, and the complex frame makes them more difficult to construct than either dugouts or rafts. Large skin boats are capable of ocean travel. The strength of these vessels comes from the frame, rather than the skin.

Bark boats. Bark boats consist of layers of a suitable bark (such as birch) covering a light frame. Native American canoes are fine examples of this type of vessel. Bark boats are limited in size and are frail compared with dugouts or rafts. They are suitable for travel only over lakes or along

*Freeboard is the vertical distance from waterline to the top of the hull, or the deck, of a boat.

inland waterways and represent an evolutionary dead end. In contrast to skin boats, bark boats derive their strength from the external shell, rather than from the frame. This distinction leads to bark boats being classified separately from skin boats.

Of the primitive watercraft developed in prehistory, the dugout and the raft had the greatest potential for further development.

South Pacific Tradition

Catamarans have been around for at least three millennia. The traditional cats consisted of a double hull of logs connected by a rigid deck or platform. Westerners first learned of cats from an English adventurer in India, who wrote about them in the 1690s. The word "catamaran" is Tamil, and these vessels have long been associated with the state of Tamil Nadu in southern India; cats have been recorded there since the fifth century CE. The cats that we know today more closely resemble those developed by Polynesian people, who spread across Oceania on large ocean-going catamarans from their origins in Asia or New Guinea about 3,000 years ago.

The vessels built by these intrepid colonizers were 50–60 ft long, each with a deck of crossbeams that was lashed to the two hulls, and supported perhaps 25 people (more, for the largest vessels), plus provisions. The Polynesians spread eastward to the Solomon Islands, on to Fiji, Tonga, and Samoa, and then to the Cook, Society, and Marquesas groups of islands. From these pinpoints of land in the South Pacific the Polynesians radiated out to New Zealand, Easter Island, and Hawaii, ending their long journeys in about 1000 CE. We have a good idea what these ancient catamarans looked like: see figure 1.1. There is still some disagreement among historians as to whether the Polynesians knew where they were going. Did they navigate to known destinations, or were the island groups they settled hit upon by accident? Whichever, it is clear that the Polynesian sailors and navigators knew how to cross open ocean in open vessels. They were capable of traveling from Tahiti to Hawaii, a distance of 2,000 miles. They navigated by stars and by birds.* Their

*Also by cloud formations, by currents, and (so Hawaiian local tradition has it) by observing the night-time glow of erupting volcanoes.

Figure 1.1. Model of a Hawaiian catamaran of 500 CE. Note the steering oars and unusual sails. Thanks to Steve Priske for providing this image.

vessels had sails made of matting and a steering paddle, and could sail well across wind and downwind. It is also likely that the Polynesian catamaran could make its way a little to windward.

The catamarans that Westerners discovered on the Indian coast and in the South Pacific were highly evolved ocean-going craft. In the 1870s they inspired Nathanael Herreshoff, a well-known and talented American naval architect and engineer who later went on to design a string of early America's Cup–winning boats. Herreshoff built cats that made people sit up and take notice, and so started the modern adaptation of these sleek vessels as pleasure craft. Catamarans and monohull yachts may nowadays occupy adjacent berths in marinas all over the world, but it is interesting that these two types of modern pleasure craft have such different lineages. We will see later why catamarans and yachts perform so differently.

East Asian Tradition

The name *junk* (from the Malay word *djong*, meaning "boat") is applied to a wide variety of ocean-going sailing vessels from India and, in particular, from China. At their peak in the early fifteenth century, junks were by far the largest and the best sailing craft in the world. Junks maintained pole position until the nineteenth century, when ships from the Atlantic tradition eventually overhauled them in terms of performance and matched them in terms of size.

The main characteristic of the junk is that it is a flat-bottomed boat without a true keel, with internal bulkheads that divide up the ship, both along the length and across the beam, into twelve or more separate watertight sections.* The hulls of large Ming dynasty ships were built in sections, which were then bolted together with large brass pins. These hulls were made of three layers of hardwood built around a teak frame. External planks were caulked with coir and then lacquered. The decks were covered; the stern deck was raised to provide protection in a following wind. Because of the flat bottom, a junk required a great deal of ballast for stability. The prow was blunt and the stern flat, and with its broad beam, the junk was rather ungainly looking to Western eyes. However, the internal bulkheads and stubby shape made the hull structurally rigid. In waters frequently swept by typhoons, strong hulls were essential. And just in case a hull compartment was holed, the junks were equipped with bilge pumps.

Large junks possessed at least four masts, each with an elliptical lugsail hanging obliquely from a yardarm. These sails were aerodynamically shaped, formed of narrow panels that were stiffened by relatively light but strong bamboo battens (fig. 1.2), which considerably simplified the required rigging compared with that of Western sailing ships. The battens were tied at each end to a sheet (a line) so that the wind force would be borne by the sheets as well as by the masts. A ship's crew could furl or reef these sails like Venetian blinds, quickly and efficiently. Junks

*This feature of junks was known to the West by the eighteenth century. Ben Franklin proposed that it would be a good idea if American ships were similarly compartmentalized.

Figure 1.2. Top: A small modern junk in Hong Kong. Note the high stern and the sail battens with simplified rigging. I am grateful to Sanjay Pindiyath for this image. *Bottom:* Illustration from an 1848 London newspaper of a visiting Chinese junk, the *Keying.*

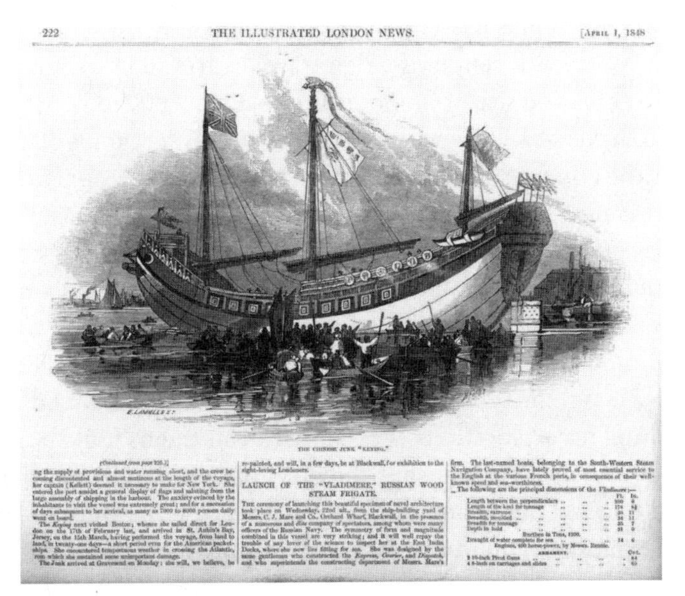

222 THE ILLUSTRATED LONDON NEWS. [April 1, 1848

THE CHINESE JUNK "KEYING."

[Continued from page 220.]

ng the supply of provisions and water running short, and the crew becoming discontented and almost mutinous at the length of the voyage, her captain (Kellett) deemed it necessary to make for New York. She entered the port amidst a general display of flags and saluting from the large assembly of shipping in the harbour. The anxiety evinced by the inhabitants to visit the vessel was extremely great; and for a succession of days subsequent to her arrival, as many as 7000 to 8000 persons daily went on board.

The *Keying* next visited Boston; whence she sailed direct for London on the 17th of February last, and arrived in St. Aubin's Bay, Jersey, on the 15th March, having performed the voyage, from land to land, in twenty-one days—a short period even for the American packet-ships. She encountered tempestuous weather in crossing the Atlantic, from which she sustained some unimportant damage.

The Junk arrived at Gravesend on Monday: she will, we believe, be re-painted, and will, in a few days, be at Blackwall, for exhibition to the sight-loving Londoners.

LAUNCH OF THE "VLADIMERE," RUSSIAN WOOD STEAM FRIGATE.

The ceremony of launching this beautiful specimen of naval architecture took place on Wednesday, 22nd ult., from the ship-building yard of Messrs. C. J. Mare and Co., Orchard Wharf, Blackwall, in the presence of a numerous and *élite* company of spectators, among whom were many officers of the Russian Navy. The symmetry of form and magnitude combined in this vessel are very striking; and it will well repay the trouble of any lover of the science to inspect her at the East India Docks, where she now lies fitting for sea. She was designed by the same gentleman who constructed the *Express, Courier,* and *Dispatch,* and who superintends the constructing department of Messrs. Mare's firm. The last-named boats, belonging to the South-Western Steam Navigation Company, have lately proved of most essential service to the English at the various French ports, in consequence of their well-known speed and sea-worthiness.

The following are the principal dimensions of the *Vladimere*:—

	Ft.	In.
Length between the perpendiculars	[illegible]	[illegible]
Length of the keel for tonnage	178	[illegible]
Breadth, extreme	[illegible]	11
Breadth, moulded	[illegible]	11
Breadth for tonnage	[illegible]	7
Depth in hold	[illegible]	[illegible]
Burthen in Tons, [illegible]		
Draught of water complete for sea	14	6
Engines, [illegible] horse-power, by Messrs. Rennie.		

ARMAMENT.

	Cwt.
2 10-inch Pivot Guns	[illegible]
4 8-inch on carriages and slides	[illegible]

handled well with the wind and on a broad reach, and even a point or two to windward.*

Sturdy and efficient Chinese junks were crossing oceans in the second century CE. Their characteristic lugsails were developed by the third century, by which time ocean-going vessels were traveling from the Yellow Sea to the Red Sea, carrying 700 people and perhaps 260 tons of cargo. These early ships were constructed mainly from softwood; their sturdiness arose from the internal bulkheads (which also, of course,

*The nautical compass is divided into 32 *points;* thus, one point corresponds to 11°15′, i.e., to 11¼°, as shown in fig. I.1.

served to reduce flooding when the hull was holed). By the eighth century, junks sported leeboards and centerboards to improve stability. Early on they were provided with large stern-mounted rudders. These rudders were mounted centrally and extended along a watertight shaft that passed through the deck and bottom. Much impetus was given to further junk development during the Song dynasty (tenth through thirteenth centuries) when, to compensate for the loss of their northern empire, the Chinese decided to expand overseas trade. Such trade required large fleets of ocean-going merchant vessels and of naval escorts to protect them. Westerners traveling to the Orient reported seeing commercial junks of typically 200 to 800 tons displacement, carrying about 130 sailors, plus passengers.

By the fourteenth century foreign travelers to China were noting the large and increasing size of Chinese junks—up to 2,000 tons. Marco Polo claimed to have seen junks with as many as 60 private cabins on the main deck, for the comfort of merchants.* The largest vessels appeared in the following century. These were the giants of the Ming dynasty treasure fleet. The capital ships measured 120 m (400 ft) from stem to stern, with a displacement of up to 3,400 tons. These behemoths were six times the length of contemporaneous European caravels. In a brief and uncharacteristic period of imperial Chinese curiosity about the outside world, Admiral Zheng He took his large fleet of gigantic junks on a series of wide-ranging explorations (I hesitate to call them "junkets"), until a change of regime slammed shut the door onto the outside world. The fleets were called home and dismantled in the 1420s. During this brief outward explosion, Chinese junks traveled southeast and down the coast of East Africa, and south as far as modern Indonesia. Menzies claims controversially that Zheng He's fleets wandered much further afield than this, reaching the coasts of Australia, West Africa, South and North America, Greenland, and the Arctic Ocean. Whether or not the junks achieved this circumnavigation of the world a century before it was accomplished by Europeans, these vessels were certainly capable of such journeys. In 1848 CE a Qing dynasty junk rounded the Cape of Good

*The claims of large sizes for Chinese junks have been confirmed by marine archaeology. In 1973 a large junk, constructed in part from cedarwood, was discovered during dredging operations off the southeast coast of China. The junk was a merchantman dated to about 1274 CE.

Hope and docked in the United States and in England (fig. 1.2). It was praised at the time by Western observers for its seaworthiness and speed (Boston to England in 21 days).

Finally, there is one unique (to the best of my knowledge) and clever feature of these historical junks that merits special mention. The bow section contained a compartment that was intentionally leaky. Water could enter this compartment and then drain away as the junk pitched in heavy seas. This design ameliorated the pitching motion, and it worked as follows. When the bow plunged down into the sea, the air-filled compartment provided a "righting moment" that reduced the downward pitching. The compartment then filled with water. As the prow pitched upward, the water-filled compartment provided a force acting downward, to again counter the pitching motion. By the time the prow again plunged into the water, the compartment would have drained. Thus, the leaky compartment was mostly filled with water when high above the waterline and was largely empty when beneath the waterline, with beneficial effects on pitching motion. Careful design must have been required to ensure that the compartment drained and filled at just the right rate.

European Antiquity

Now we must take a giant leap backwards from the imposing Chinese junk to the origins of the Atlantic tradition of ship-building, before taking many short and stuttering steps—and a few large strides—forward. We will reach (at the end of chapter 3) what were arguably the best and most beautiful of historical sailing ships: the full-rigged clippers of the mid-nineteenth century.

There were two independent strands of ship-building development in Europe, which we might classify with equal validity as "North Sea" and "Mediterranean," or as "clinker-built" and "carvel-built." I begin by describing each strand separately, before showing how their fusion in the Middle Ages led to a rapid advancement in nautical technology that can aptly be described as a revolution.*

The terms *clinker-built* and *carvel-built* refer to the manner in which hull planking is put together. This seemingly arcane distinction is, in

*OK, so it was a revolution *adagio*, in slow time, compared with the *allegro* tempo of the fast-paced industrial revolution that changed the world in the late eighteenth and early nineteenth centuries.

fact, crucial to ship development, as we will see, and the two terms reflect the technological state of the societies that gave rise to them.

Egyptians and Phoenicians

You may wonder with some justification how I can possibly consider the ancient Egyptians and the Phoenicians to be part of European antiquity, given that these nation-states were located in North Africa and the Middle East. Well, I may be stretching a geographical point here, but in terms of maritime evolution the Egyptians and Phoenicians contributed to the Atlantic tradition, and not to any of the others, given that they sailed their ships in the Mediterranean Sea and in the Atlantic Ocean.

The earliest representation of an Egyptian boat under sail comes from a vase that has been dated to 3500 BCE. Certainly by this time there were reed rafts, each perhaps with a mast, working the river Nile. This river was of central importance to the civilization of ancient Egypt and almost certainly spurred the development of Egyptian ships. The Cheops ship, a well-preserved archeological find dating from 2600 BCE, was a funeral ship buried with the pharaoh Cheops. Excavated in 1954 from the Great Pyramid at Giza, it is 125 ft long with a hull made of wooden planks. The Cheops ship cannot be regarded as a practical vessel, of course, but it does indicate the ship-building know-how of the Egyptians, and historians have learned a great deal by studying it in detail.

By 2500 BCE the Egyptians made wooden boats, mimicking the shape of river rafts, which transported goods up and down the Nile and along the coastline of the eastern Mediterranean. These boats were flat-bottomed and square-ended, without a keel. The lack of a keel, as we have seen, suggests that they evolved from rafts. It also limited the size of mast, and hence of sail, that the boat could carry. Thus, we believe that such boats were suitable only for moving up and down the coasts and not for open ocean travel. More barge than boat, this type of vessel (for which there is a lot of archaeological evidence) could be as much as 200 ft long and 70 ft across the beam.

The Egyptians also built seagoing boats or barges, but the seagoing vessels seem to have been little more than reinforced versions of the river boats and coastal traders. Their hulls were made of short sections of wooden planks, fitted closely end to end, carvel style (of which more shortly). They carried a single square sail and were steered by two oars on the stern. Such vessels would have sailed well enough with the wind, and on a beam reach, but would not have been able to sail to windward.

The Phoenicians were a trading civilization spread far and wide across the Mediterranean but originating from the area around modern Lebanon. They have left us much less evidence about their ship-building capabilities than the Egyptians, although we do know that their trading vessels had sturdy hulls and decks built of wooden planks. Nevertheless, it is quite clear that the Phoenicians were the maritime experts of European antiquity. They traveled long distances by sea in order to establish trading links with other peoples. There is evidence of Phoenician ships trading for tin in Cornwall, on the southwest tip of England. Phoenicians were well-established mariners by 1000 BCE and are known to have sailed as far as the Cape Verde Islands off the west coast of Africa by 460 BCE—a journey that took them 2,000 miles from home and into the Atlantic Ocean.

In addition to trading vessels the Phoenicians also developed war galleys, prompted by trading competition with the rising stars of Mediterranean antiquity, the Greeks. Which neatly leads me to our next topic.

Greeks and Romans

The speed of a boat in ancient times was determined by the number of oars that it carried or by the spread of canvas. Both the Greeks and the Romans made use of sailing vessels for commerce but used oar-powered galleys for maritime warfare. It is the slender and beautiful galleys that have come down to us as the maritime icons of European antiquity because they were much more impressive vessels than the slower, ungainly-looking trading ships. The rise of galleys is a testament to the weakness of sailing know-how at this period of the Atlantic tradition: rigging consisted of one or two masts, each carrying a single square sail hung from a horizontal yard. With a following wind the yard was angled perpendicular to the longitudinal axis of the ship—the origin of the much more efficient square-riggers of later centuries. For sailing to other points with the wind (against the wind was not feasible in antiquity) it was necessary to twist the yard about the mast so that the sail would catch the wind at an optimum angle. I analyze the effectiveness of square-rigged ships in detail in the next chapter, but here it is necessary only to point out that sails took up a lot of deck space. The wind might come from directly behind, or from over the left shoulder or right shoulder, and the yard had to be oriented correctly to catch the wind. The sail and rigging would occupy much of the port or starboard side of the deck amidships, depending on orientation. So the midship deck area had to be kept clear for the sail. Plus, reefing the sail required a large crew. These realities caused problems for warships, which needed soldiers on deck to repel boarders or to board enemy ships. Hence, oars were preferred for warships because they did not waste deck space.

Before the days of gunpowder and cannons, naval warfare was simply land war at sea, meaning that warships were just fighting platforms for soldiers. Greek (and Roman) galleys would ram each other;* if this did not sink the enemy vessel, then soldiers poured over the gunwales to fight it out hand to hand. Ramming required speed and rapid maneuverability, and hand-to-hand fighting required soldiers; both these factors led to galleys rather than sailing ships for Greek and Roman warfare at sea.

The need for speed meant more oars, which in turn meant that galleys had to be longer than sailing ships. The typical Greek merchant ship (whose design was probably influenced by the Phoenicians) had a ratio of length to beam of 4:1, whereas the ratio for the war galley was closer to 10:1.† The Greeks themselves referred to their merchant ships, which were sailing vessels, as "round ships." An arms race started up: faster enemy galleys meant that your galleys had to be faster still to avoid being rammed and to outmaneuver the enemy. So galleys became longer and longer to accommodate more oars, for more speed. But longer ships are structurally weaker; they were at risk of *hogging,* or breaking up between heavy waves. The practical limit for length appears to have been about 95 ft, though some of the later—and presumably stronger—galleys may have reached 125 ft. When still more speed was needed, the solution was to build a second bank of oarsmen, resulting in the Greek biremes (the Roman equivalent is shown in fig. 1.3). The demand for yet more speed led to a third bank and triremes. These large Greek galleys, with up to

*Large bronze-encased rams were an integral part of the prow of Greek galleys by the ninth century BCE.

†The large ratio of galley length to beam was due to the narrow beam as much as to the extended length. A narrow beam reduces hydrodynamic drag, as we will see in a later chapter, and so helped the oarsmen squeeze a little more speed out of their vessel.

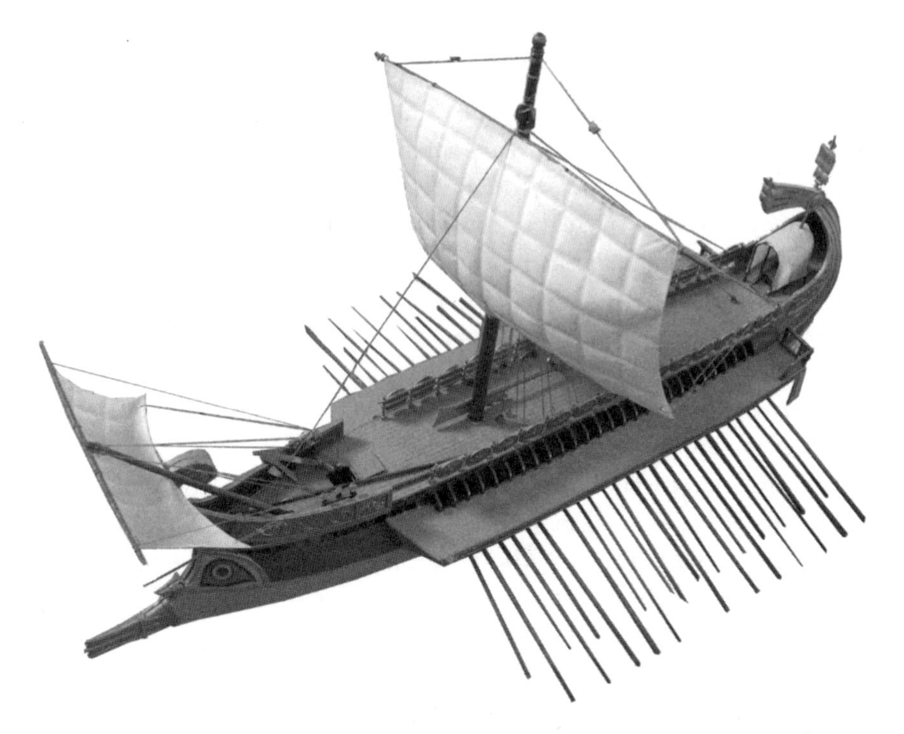

Figure 1.3. A Roman bireme with its two banks of oars on each side. Note the large square mainsail, typical of classical antiquity, though of lesser importance than the oars on warships such as this. Note also the ram attached to the prow below the waterline. Image from Wikipedia.

170 oarsmen each, dominated warfare in the Mediterranean from the sixth to the fourth century BCE. It must have taken considerable training for such a large crew to row in unison, in particular when maneuvering in the heat of battle. It was important for war galleys to have effective steering so that they could maneuver to intercept an enemy vessel and ram it, or avoid being rammed. Historical evidence and modern reconstructions have shown that the steering oars used on Greek and Roman galleys were effective because they were symmetrical, with one on either side of the stern.* Without such symmetry, as we will see, the steering oar is much less effective than the sternpost rudder.

*There is a lot of archaeological evidence in the form of shipwrecks, from which historians have learned a lot about Greek and Roman galleys and sailing ships. Also, galleys are a common motif on Greek pottery that has survived through to the present day.

Rome succeeded ancient Greece as the dominant Mediterranean power, and Roman galleys grew into the larger quadriremes and quinquiremes. Here, *quad-* and *quin-* do not refer to four and five banks of oars, but instead to four or five oarsmen per oar. The oars needed to be very long to reach the water from the top bank of a large trireme, and so more oarsmen were needed to row effectively. The Romans, unlike the Greeks, were not a seafaring people and made few nautical innovations. However, the Romans were excellent builders, and they were happy to borrow designs from subject peoples. The large Roman galleys were based on the older Greek design, with influences from Hannibal's Carthaginians (successors to the Phoenicians), and they dominated the Mediterranean Sea until the fifth century CE. These galleys were troop transporters and weapons platforms that could carry 120 soldiers plus oarsmen. The soldiers disembarked via a hinged gangplank at the stem, much like they did from a World War II landing craft.

Galleys—both Greek and Roman—would have a simple square sail for crossing the sea, to relieve the oarsmen, but sails were never used in battle. Indeed, oared vessels could not carry enough provisions for the large crew to withstand a long sea voyage, so galleys were not suitable for the open ocean. Trading over long distances required more seaworthy ships, and so the Greek "round ships" and the vast number of Roman merchant ships were sailing vessels. What were they like?

There is less representation of Greek merchant ships in art compared with depictions of biremes and triremes, perhaps reflecting their lower status. War was a noble art, whereas trade was for the hoi polloi. We have seen that the shapes of sailing ships and galleys were different because of the conflicting design requirements. In classical times the trading ships became quite large because the Mediterranean was relatively free of pirates and because the large empire (in the case of the Romans) required bulk transportation of goods—in particular, grain from Egypt to Rome.* Merchant ships of the fifth century BCE were

*That there were fewer political borders to cross 2,000 years ago than at any time since must have facilitated trade. This is because there existed three large empires at the same time—Rome, Parthia (Persia), and China—connected by a trade routes over land, the "Silk Road." Thus, silk was available to wealthy Romans, and trade in general was more widespread than later, during the European Dark Ages, when fractured empires and small states in constant flux deterred regular trade over long distances. While the Roman Empire lasted, the Mediterranean was a Roman lake,

typically of 150 tons' cargo capacity, though 300–500 tons was not uncommon. The smaller vessels were about 60 ft long and a fat 20–25 ft across the beam. By the first century BCE grain ships (*corbita*) with capacities of 1,300 tons could be found sailing the grain run from Alexandria to Rome (fig. 1.4). Grain was transported in sacks, and liquids in large *amphorae*—earthenware pots that were packed into place in the cargo hold with brushwood.

We know from marine archaeology that the hulls of Roman merchant ships were largely unchanged from those of the Greeks: large, carvel-built, and high-sided. As with the galleys, merchant ships were maneuvered via symmetrical twin steering oars that were "in no way inferior to the Medieval stern rudder," in the words of historian J. G. Landels (see also Lionel Casson). These oars were often boxed in with additional hull planking for protection.

There must have been maintenance problems for the large ships because dry-dock facilities would not have existed to take them. The shipworm *Teredo navales* must have chomped its way through a number of Greek and Roman hulls, despite the application of pitch or lead tiles below the waterline. Careening (tilting a ship at a steep angle, so that one side of the hull is exposed for cleaning and repair) was apparently unknown in classical times.

The anchors for these merchant ships (of stone, or of lead encased in wood) were correspondingly large. We know this because, being robust, many of them have survived to the present day. A typical 230-ton boat was equipped with up to five anchors, the largest being 8 ft long and weighing 1,500 lb.

Romans improved the rigging somewhat over their Greek predecessors. Blocks, pulleys, halyards, stays, sheets—most of the standard items in later sailing ships' standing and running rigging were present in Roman merchant vessels. The two or three masts carried square-rigged sails. Sails could be at least partially furled, quickly when required, from the deck, without any of the crew needing to go aloft. By the fifth century CE Roman merchant ships dominated Mediterranean trade just as their galleys dominated militarily. By this time the large square sails had been supplemented by additional sails: bowsprit sails, topsails, aft spritsails,

---

and sea trade was safe enough for regular routes to be established. Rome imported about 150,000 tons of grain from Egypt each year.

Figure 1.4. Two views of a Roman corbita (grain ship). These carvel-built merchant vessels were propelled by sails and not oars; consequently, their length-to-beam ratio was much smaller than that of the warships. Note the boxed-in symmetrical steering oars and the bowsprit sail. The swan neck decoration appears on all corbita images. Thanks to Antonis Kotzias for providing these images.

and even the new lateen rig (of which more later). These merchantmen could probably manage 4–5 knots in a good following wind. They could sail effectively with the wind on either beam (though more slowly, and with considerable leeway). The mainsail yard would be braced aslant to the wind, with the sheet on the weather side let out and the lee-side sheet tightened by a winch. Apparently one well-known problem at the time was the tendency for the ship to "head up"—turn to windward when the wind is on a beam.* It was difficult to counter this tendency in a large merchant ship via the steering oars alone. (In this respect steering oars were less effective than the later sternpost rudders.) Instead, the mainsail was shortened on the lee side.

It is likely that these ships could sail a point into the wind, but no more. We will see that later square-rigged sailing ships could sail at least two points into the wind. This difference may not sound like much, but it is very significant. When the wind is dead ahead, tacking (discussed in detail later) is laborious and slow. Landels notes (correctly) that, had the ancients been capable of improving their ships to sail even 1° closer to the wind, they would have reduced the distance traveled by 8%. By sailing two points into the wind, the square-riggers of the Age of Sail reduced the distance traveled by nearly half.[2] We can appreciate the exertion required for Roman sailing ships to sail into the wind from the following record. It is reported that merchant ships plying a regular route from the Straits of Messina in southern Italy to Alexandria in Egypt required 18–20 days to make the journey. The dominant wind is from the northwest, and so the outward journey was largely with the wind. The return journey, on the other hand, required between 40 and 65 days.

Despite their clumsiness and their lack of "Wow!" compared with the contemporary war galleys, the workaday Roman grain ship contributed significantly more to the development of Mediterranean ships.

Carvel versus Clinker

The front-cover difference between clinker-built hulls and carvel-built hulls is illustrated in figure 1.5. The front cover does not represent the whole book, however, but like hull planks, or *strakes,* is what you see on the outside. Carvel-building probably came first† and originated in the

*The reason for this behavior is discussed in chapter 5.

†Carvel-built vessels were standard for the first millennium BCE in the Mediterra-

eastern Mediterranean region. It was a natural extension of the ancient Egyptian way of building barges, with short planks fitted edge-to-edge, and seems like the natural way to construct a hull. Carvel planking was long, perhaps extending the length of the hull in small ships. If the planks were not as long as the hull, they were butted end to end. Planking was sawn to shape, and the planks were fitted snugly side by side. The strakes of early ships were held in place by ropes that passed through holes (stitched together, as Homer put it), but later, strakes were attached to each other via the mortise and tenon method (see fig. 1.5). The key point here is that the main strength of the boat came from the frame, and not from the manner in which strakes were connected to each other. To construct the boat, first the keel was laid down, then the stem and stern posts and transoms were attached to it, and then the ribs and the rest of the frame—the ship's skeleton—was fitted. Once the frame was complete,* then planking was fixed to it, following the line of the keel and stem / stern posts (so that the strakes rose up at prow and stern). Once one strake was in place, the next strake was fitted to it and fixed to the frame.

Carvel-built ships were more streamlined than clinker-built boats because the hull was smoother. They required more caulking, however, to make them watertight and were more labor-intensive. The main advantage of carvel-built ships, which led in time to the spread of carvel-building techniques across Europe, was that they could be bigger than clinker-built boats. As we will see, there was a fundamental limit to the size of clinker-built vessels; no such limit applied for carvel-built ships. Also, carvel planking could be thicker, and therefore stronger, because it did not have to be bent into position. (In many parts of the world, clinker planking is called *lapstrake*, because the strakes overlap.)

Clinker-built boats were made in northern Europe. The boat-builders in these wilder and less civilized fringes possessed no saws, or no saws that were precise enough to make squared-off, close-fitting planks. One advantage of clinker-building was that it required only adzes. First the keel, stem, and stern posts would be laid down, and then rough-hewn

---

nean. The first clinker-built boat known to archaeology is the so-called Nydam ship; it is dated to 350 CE and is about 82 ft long.

*To be pedantic, sometimes only that section of the frame that lay below the waterline was completed before planking was attached, with the rest of the frame put in place afterwards.

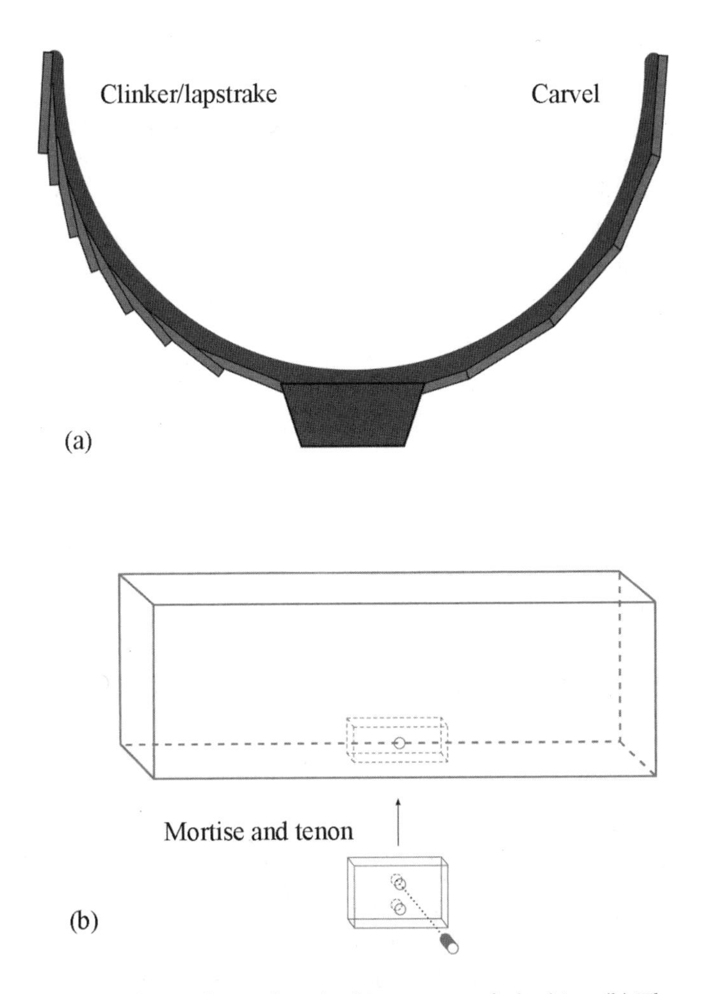

Figure 1.5. (a) Clinker, or lapstrake, planking vs. carvel planking. (b) The mortise and tenon joint connecting two carvel planks. The main strength of carvel boats, however, came from the planking attachment to the frame. For clinker hulls, the strength was in the shell. Carvel hulls created less drag, but were more difficult to construct.

planks (riven timber, or split wood) would be attached. These planks ran from stem to stern, were bent into place, and then were nailed flush to the posts. Boat size was limited by the fact that the planks had to run the length of the hull.* The overlapping planks were fixed to each other by

*Consequently, clinker building continued until a much later date for boats than for ships.

nails that were driven through the overlapping sections from the outside, and then turned back on themselves (*clinched*, hence "clinker") to form a hook shape, before being driven back into the inside plank. Rivets achieved the same task more effectively in later boats. For larger clinker boats with thick planking, oak spikes called *treenails* (pronounced "trennels") were driven into undersized holes to hold the strakes together.

The strength of clinker-built boats lay in the outer shell of planking. The internal framing timber was added after the outer shell (bent into shape by steaming) had been built. This contrast between carvel and clinker vessels echoes the prehistoric difference between skin and bark boats. In one case the hull strength resides in the internal frame; in the other it arises from the outer shell.

Viking Longships

I will take the justifiably famous Viking longship as the apogee of the northern European clinker-built strand of the Atlantic tradition. Of course there were other types of clinker-built ships, but discussing all of these would risk turning this chapter into an encyclopedia of nautical evolution. I will here utilize the Viking longship as a metaphorical clothesline on which to hang the showy outer garments of clinker-built technology; should you be inclined to delve into the many and varied types of nautical underwear, then you are on your own.

The Viking longship represents the absolute maximum size for a clinker-built vessel. The long planks were attached to a large, strong keel that was T-shaped in cross section. These double-ended boats were designed with low freeboard because oars provided the main source of power, especially in the early days. As with Greek and Roman vessels (though on a smaller scale), the Vikings relied on oars (perhaps 80 of them) for their warships and sails for their merchant craft (the high-sided knorr). By 800 CE, though, the longship had a large square sail set on a removable mast that was deployed for sea voyages. Given the low freeboard, it seems unlikely that the sail would have been trimmed for anything but a following wind. By 1000 CE longships and their intrepid crews had spread from Norse countries to Iceland, Greenland, and even Newfoundland. Such voyages across turbulent seas are impressive, considering that the longship is an open boat with no deck, and is a testament to the strength of the clinker hulls as well as to the people aboard them. The Swedish contingent of Vikings took their shallow-draft longships up rivers deep into Russia and as far afield as the Black Sea. The shallow draft of longships was partly a consequence of their northern origins: clinker boats had a lighter frame and so sat higher in the water. To some extent this countered their greater hydrodynamic drag, compared with carvel vessels.

Unlike Mediterranean seafarers, the northern boat-builders had not discovered the benefits of symmetrically placed steering oars, and so steering longships was not as easy as steering Roman galleys. By convention, the single oar was positioned on the right side of the boat, and so the right side (the "steerboard") became known as the starboard side. When the merchant knorrs were approaching a quay for loading cargo, they had to keep the steering oar away from the quay so that it would not get crushed and so that the gap between the gunwale and the quay would be minimized. Thus, the left side of the boat was used for loading—it was the "lading board," or larboard.

Many people (particularly in Scandinavia) would argue that Viking longboats were the most beautiful boats produced in the first millennium CE. Numerous archaeological remains have been found from different centuries. Among these are the Oseberg ship, retrieved in 1904 from a ninth-century burial mound in Norway, and the tenth-century Gokstad ship, disinterred in 1880 from another burial site, also in Norway (fig. 1.6). Both these vessels now have their own websites, should you be interested in seeing them (see bibliography). They were between 70 ft and 80 ft long, with room for about 30 oars; the Gokstad ship had a 17-ft beam. Both were steered by a single steering oar controlled by a tiller. Replicas of these vessels have crossed the Atlantic several times—a tribute to their seaworthiness.

The Middle Ages: Technological Revolution

If you are a history buff, it may surprise you to hear the words "Middle Ages" next to the words "technological revolution." After all, the Middle Ages in Europe were times of social stagnation in many ways. People's lives did not change much from one generation to the next; the son pursued the same craft as his father, and the vassal owed liege to his lord, down many generations. On land, indeed, life stumbled on unchanged by much in the way of technological advances during this period. At sea, however, it was a different matter entirely.

Figure 1.6. Right: A model of the Gokstad longship. A clinker-built boat, with no decking and open to the elements, this and other Viking warships were characterized by low freeboard to facilitate rowing. Longships were something of a throwback to older times in terms of propulsion (with a square sail playing second fiddle to the oarsmen), but the hull construction represented the epitome of clinker building. I am grateful to Slava Petrov for this image. *Left:* Photograph of the Oseberg longship. Image from Wikipedia.

The Medieval Contribution

We have now arrived at the start of the second millennium CE. The Roman Empire has come and gone—but its dissolution provided a spur to sailing ship development. This is because the political turmoil that followed in the wake of Rome's demise resulted in more fluid borders and changing trade routes. The predictable year-in, year-out voyages of merchantmen between port cities of the Empire disappeared. Consequently, merchant ships needed to be able to tackle a wider variety of wind and sea conditions, and to travel further afield with more varied types of trade goods on board.

We have seen that the Atlantic tradition of ship-building consists of two strands, carvel and clinker, that differ in the manner of hull construction. Oar-powered galleys have given way to sailing vessels, mostly square-rigged. Rigging has become, by the year 1000 CE, fairly well advanced, and anchors are similar to modern ones, with recognizable shank and flukes.

A big step forward had occurred a few centuries earlier with the introduction of the lateen sail (fig. 1.7). This large triangular sail hung from a very long, oblique yard, suspended from the mast in the middle, and probably was introduced to the Mediterranean region via the Arab dhow.* The Byzantine Empire, successor to the Roman Empire in the East, with its capital at Constantinople, fought the Arabs; and both sides made use of warships that we know as *dromons*. These ships were hybrids, halfway between the war galleys of antiquity and the ships that spread Europeans around the world during the Age of Exploration. The dromon was oar-powered like the galleys, but unlike them it possessed a significant amount of canvas, in the form of two large lateen sails hung from two masts. Wind power was as important as oar propulsion.

These vessels must have been quite fast and maneuverable. The dromon varied in length between about 100 ft and 160 ft—quite long, we recall from our earlier consideration of Greek and Roman galleys, so as to accommodate the oarsmen (one or two banks). The lateen sail required a lot of space (and crew, because the long yard was heavy and difficult to move), and the long hulls of the dromons provided this space. Lateen sails were (and still are†) fore-and-aft sails, hung along the longitudinal axis of the ship and thus able to take the wind on either side, and so they improve maneuverability and permit sailing closer to the wind, as we will see later. Dromons were about 20–25 ft wide, so that the ratio of length to beam is about 5:1 or 6:1, much less than in the old galleys. Such width would have been necessary for stability because the large sails would cause the ship to heel in a crosswind. The large deck area of dromon warships was taken up by numerous catapults, some of them throwing Greek fire (an early version of napalm), and by archers. Note the shift of emphasis from ramming and boarding—land battles at sea—to more modern stand-off battles, with combatants bombarding each other from a distance.

*It is thought that the lateen sail first appeared in the eastern Mediterranean in the second century CE, introduced there (though widely adopted only much later) via Egypt or the Persian Gulf.

†The modern Sunfish class of one-design boats are lateen-rigged.

Figure 1.7. The lateen sail has a very long, canted yard that reaches down almost to the deck. The oldest type of fore-and-aft sail, it provided medieval European ships with some ability to sail into the wind. The large yard required a lot of clear deck space and a large crew. Thanks to Simona Manca for providing this image.

The combination of oars and lateen sails was the key to dromon success. These ships were widespread within Byzantium and were used as supply vessels and escorts as well as warships. The dromon ruled the eastern Mediterranean from the time of the great Byzantine emperor Justinian in the sixth century to the twelfth century CE, during which period the old square sail disappeared almost completely from the region. The square sail remained important in the rest of Europe, however —in the "clinker zone" to the north and west—because of the rougher seas. The new-fangled lateen sail ("latin sail," to the northerners) worked better in the sheltered Mediterranean Sea but worse in the stormy North Sea and Atlantic Ocean. On the other hand, the lateen sail permitted tacking into the wind, whereas the old square sails could only move a ship downwind or, at most, across the wind (a beam reach).

A second big step forward occurred toward the end of the medieval period. From the twelfth century CE, boats with stern rudders appeared in the cold waters around Scandinavia. The stern rudder is more effective at steering a boat than is a single steering oar, and so use of the stern rudder spread from the Baltic Sea across northern Europe (probably dispersed through German sea traders). The stern rudder had been invented in China during the eighth century CE or earlier, but of course Chinese technology was unknown in the West at this time.*

Mediterranean ship-builders remained ignorant of this innovation for a while longer, but their need for stern rudders was not so great because the twin steering oars on Mediterranean ships were almost as effective as rudders. Almost but not quite: the stern rudder is fixed to the stern post and so is held more firmly to the hull and can be made much bigger. Nobody in Europe knew it yet, but the potential of stern rudders extended far beyond improved steering, important though that was. Because stern rudders can be large, they can control larger ships. The use of steering oars limited the size of Mediterranean-built ships, particularly in the open ocean, where greater freeboard was needed, because there is a practical limit to the length of steering oars if they are to be effective. In the north of Europe this constraint on ship size was removed with the adoption of stern rudders, but as we know, the northern clinker or lapstrake hulls were limited in size for other reasons.

So long as the clinker and carvel strands of Atlantic tradition ships remained separate from each other, each remained ignorant of the advantages possessed by the other. It was the combination of the two strands that proved to be explosive and led to a rapid growth in the size of sailing ships.

Cogs in the Wheel of Commerce

One of the earliest European ship designs to adopt the centrally mounted stern rudder, controlled by a tiller, was the cog, the workhorse of the Baltic and North Sea trade from the thirteenth to the fifteenth centuries CE. The cog replaced the old Norse knorr, which had been adequate for economically undeveloped societies but not for the international trade

*We know when stern rudders took over from steering oars in northern Europe because (fortunately for historians) ports in Flanders kept records of visiting ships. Records for the year 1252 CE distinguish between ships with oars and those with rudders, and so this year was a time of transition from oars to rudders.

Figure 1.8. Left: Model of a medieval cog; note the stern rudder, and the fore and stern castles. *Right:* Barrels of cargo in the hold. Once again, I am grateful to Slava Petrov for these images.

that grew from the twelfth century onwards. This increased trade is largely associated with the northern European Hanseatic league, a group of enterprising merchants based in the major ports of northern Europe. They traded grain, herring, timber, salt, and just about any bulk produce that could be moved by boat. The cog was flat-bottomed so as to more easily navigate shallow waterways and estuaries, with a removable keel. The design details evolved, but in general these ships were high-sided and clinker-built (except for the flat bottoms). Cogs were heavily built to be rugged, and so handled clumsily, like a barge. They possessed a small raised fighting platform at bow and stern. These platforms—forerunners of the forecastles and stern castles that grew to be so prominent in sixteenth-century galleons—helped crews to fight off pirates and competitors who would disrupt trade (fig. 1.8). In addition, there was sometimes an enlarged crow's nest that provided a third fighting platform.

Whereas the knorr could carry about 50 tons of cargo, the cogs could carry between 100 and 400 tons. Stern rudders permitted this increased capacity, as well as improving ship handling. The deck of a cog consisted of heavy planking with gaps between the planks (to aid drainage). Consequently, water was frequently pumped from below deck by bilge pumps. This design choice seems strange because it must have made the transport of perishable goods such as grain more difficult. In fact, cargo was not carried loose, in bulk, but rather, in large, 250-gallon barrels called *tuns* (hence our word *tonnage*). These large barrels were loaded and unloaded via the yardarm of the main mast, used as a crane.

The typical cog had only one mast, which was square-rigged and was pretty much incapable of sailing to windward. This shortcoming did not matter so much in the North Sea and the Baltic, where wind direction changed seasonally. The tiller was often covered for the benefit of the helmsman and any passengers that might have been on board—hence our word *steerage*. We know a great deal about cogs because there are a lot of historical and archaeological remnants from 600 years ago, when these ships were at their zenith.*

The Incredible Hulk

The hulk originated in the Low Countries (modern Holland and Belgium); it is associated in particular with the Frisian Islands off the coast of Holland and northern Germany. While it was developed before the cog, it came to replace the cog only in the early fifteenth century.† The hulk had a larger, rounder hull than the cog and a much deeper, wedge-shaped cross section ending with a shallow, banana-shaped keel. Because of its deep draft, the hulk was more stable with better (i.e., less) leeway. Hulks were clinker-built, in the northern tradition, with fore and aft castles, and a top castle, like the cog. All the strakes that constituted then hulk shell terminated above the waterline, and so the hull was upturned at stem and stern, a feature that characterizes the hulk. It had a sealed deck, which seems a much more sensible construction than the leaky cog deck, in that the cargo was better protected. The hulk was sufficiently common and important to have been depicted on coins of the period. At least one of the earlier hulks (from about 800 CE) has been excavated by marine archaeologists near Utrecht.

I mentioned the castles added to defend these merchant ships. We are now in a transition period of European history. In the north, the Hundred Years War (1337–1453 CE) between England and France led to much fighting at sea and to the introduction of armed convoys of merchant

*For example, in 1962 a fourteenth-century cog was excavated in Bremen, Germany. It was 77 ft long with an estimated carrying capacity of 130 tons.

†The medieval hulk must be distinguished from the modern hulk, which today means a dismasted and damaged sailing ship; there appears to be no clear linguistic connection between the two uses of the term *hulk*.

ships. Privateers became common and would remain a factor for several centuries. Later in the fifteenth century, deck cannon were added to the merchant ships' armory, and then below-deck cannon with gun ports piercing the hull. Medieval ships were beginning to specialize, transitioning from armed merchantmen into either warships or trading/cargo vessels. (Increasing the number of guns and gun crews meant less space for cargo.)

The Age of Exploration: Carrack and Caravel

In the early fifteenth century the pace of ship development in the Atlantic tradition began to pick up: this was the period with the most intense development of sailing ships. I hinted earlier about the potential for ship growth (in size) and development should the northern clinker and the Mediterranean carvel designs fuse. Such a mixing began with the Crusades. Many northern knights headed for the Holy Land in their double-ended clinker-built ships, taking them into Mediterranean ports en route. Better ships led to increased trade further afield, which contributed to the mixing of ship-building ideas so that, instead of two separate strands, Europe became a melting pot of diverse ideas concerning ship design. One of the first fruits of this melding of ideas, and certainly the best-known and most successful ship of its day, was the carrack.

The carrack was a carvel-built ship with a stern rudder. This combination of features permitted carracks to grow larger than their predecessors. Originating in the early fourteenth century around Genoa, by the fifteenth century a typical carrack displaced 600 tons; by the sixteenth century carracks weighed in at 1,600 tons. These later carracks were typically 115 ft long with a beam of 33 ft and a depth of 17 ft. The carrack was a high-sided vessel with a stern castle and a particularly high forecastle. These castles grew into multistoried structures, giving the carrack a characteristic U-shape side on. The castles were integral to the hull and not, as earlier, appearing to be add-ons. They provided protection from the weather amidships, as well as protection from attackers. Carracks heeled rather a lot in strong winds, being somewhat top-heavy because of the castles. Their sterns were rounded, and they had a relatively deep draft: they were open ocean ships.

The early carracks were two-masted, with the main mast square-rigged and the mizzen lateen-rigged. Here is another fusion of the two European strands that yielded significant benefits: the northern square-

Figure 1.9. A model characterization of a carrack, with intentionally exaggerated features. Note the high stern castle, the stern rudder and, especially, the deep draft of this archetypal Age of Exploration ship. Thanks to Slava Petrov for these images.

rig sails were good at running before the wind, while the Mediterranean lateen sail permitted tacking and yielded better control. Later carracks had three masts, the new foremast being square-rigged. Bowsprits were added early on (around 1350) to the top of the forecastle, and by the mid-1400s some carracks had topsails above the main sails. Masts, consequently, grew taller. As the carrack evolved, a fourth mast, the bonaventure, was added aft of the mizzen and given a lateen sail. From 1500 CE all four masts carried topsails, and the main and fore masts carried topgallants. Thus, carracks were among the first fully rigged ships to be built (fig. 1.9). The proliferation of sails resulted in crew specialization, with some crew stationed in the rigging rather than, as earlier, with everyone working the sails from the deck. The carrack was the best Atlantic-tradition ship to date and was used for trade and war throughout Europe* and for exploration beyond Europe.

*Carracks first reached England in the early sixteenth century, brought there by Genoese traders. The most famous English carrack, however, suffered from being a "prestige" ship. She was the *Mary Rose*, King Henry VIII's flagship, built in 1510 and

The seaworthiness of carracks is clear from the role that they played in the Age of Exploration. Of the three ships that took Christopher Columbus to the New World in 1492, one, the *Santa Maria,* was a carrack (fig. 1.10).* In 1519 Magellan first circumnavigated the globe in the carrack *Vittoria.*† Portuguese carracks reached as far as Japan, where they were known as "black ships" because of the pitch on their sides. The carrack was the beast of burden of the Age of Exploration and became the standard vessel of Atlantic trade. As well as being an excellent transport ship, it provided a stable deck that was an excellent gun platform. So here we see two motivations for the evolving Atlantic-tradition sailing ship. At the junction in history where northern and Mediterranean strands come together, we have on the one hand the need to expand trade and explore the world, and on the other hand the need to provide effective platforms for increasingly effective (and increasingly large) cannons.‡

The other great ship of the Age of Exploration is the caravel.§ Originating in Portugal and derived from thirteenth-century fishing boats, the caravel was adopted by most European seafaring nations by the fifteenth century, though it will always be associated with Iberian explorers. Significantly smaller than the carrack (typically 65–80 ft long, 25 ft in the beam, and 10 ft deep) and with a correspondingly smaller cargo capacity (60–100 tons), the caravel was nevertheless a successful merchantman of luxury goods such as silks and spices in the Atlantic and Indian oceans

---

sunk, under the baleful eye of her king, off the southern coast of England in 1545. Her gun ports were too close to the waterline, and she shipped water when heeling during a maneuver. The *Mary Rose* was raised in 1982 and is now on public view (see bibliography).

*This is the consensus view of historians, a minority of whom consider the *Santa Maria* to be a caravel. Certainly this ship was not a standard carrack (compare fig. 1.10 with the description of carracks in the text).

†Only 18 of the original crew of 270 completed this eventful journey. Their captain, a Portuguese in the employ of the Spanish crown, was killed en route. Nevertheless, Magellan is given credit for the first circumnavigation because he had earlier been the first person to pass through every meridian. The carrack was known as the *nao* to the Portuguese and Spaniards.

‡Clinker hulls were not well suited for piercing to make gun ports because the strength of clinker-built ships lay in the shell, not the frame. This strength was compromised by piercing the hull.

§The name *caravel* is perhaps linked to *carvel,* as in "carvel-built."

Figure 1.10. The *Santa Maria*, a carrack, was one of the three ships taken by Christopher Columbus on his first voyage to the New World. This picture seems hardly less fantastic than the model of figure 1.9. Again, the high stern is prominent. Compare the increasingly complex rigging with that of earlier ships. Image from Wikipedia.

and a ship of exploration par excellence. Two of Christopher Columbus's ships were caravels: the *Niña* (fig. 1.11) and the *Pinta*. The Portuguese explorers Bartolomeu Dias and Vasco da Gama were sent to the four corners of the world by their monarch, Prince Henry "the Navigator," in caravels.

The caravel sailed better than the carrack (which by comparison was slow and ponderous, and handled poorly); it rolled less and, with better lines and consequently less drag, was faster.* Carvel-built like the carrack but with a shallower draft, the caravel hull design made the ship buoyant and resistant to leeway. It had a raised stern but no forecastle. The aft castle overhung the square stern, resulting in a poop deck. The caravel was small enough and with a low enough freeboard so that it could be rowed in a pinch. The three or four masts were usually lateen-rigged, so that caravels sailed well into the wind. (Caravels were, arguably, the best windward-sailing ships in the world during their heyday.) Some caravels were switched from lateen- to square-rigged to take advantage of a following wind; the *Niña* is one example. This differentia-

*The caravels were also more maneuverable in tight bays and close to rocky shorelines. Of Columbus's three ships only the carrack *Santa Maria* foundered.

Figure 1.11. Replica of the *Niña,* the larger of the two caravels taken by Christopher Columbus on his first voyage to the New World. Again, the high stern is prominent. Note the combination of square and lateen sails. Images used by permission of the Columbus Foundation, B.V.I.

tion of rigging led to different names: *caravela latina* if lateen-rigged for sailing closer to the wind, and *caravela redonda* if the fore and main masts were square-rigged for better all-round performance. The different names reflect the differing sailing characteristics and requirements of the two versions.

Caravels remained the Europeans' ship of choice for open ocean exploration and trade until the end of the sixteenth century. By this time, the little caravel was just too small for the more demanding needs of a changing world.

Intermission

My history of the technical evolution of Atlantic-tradition sailing ships has progressed from prehistoric dugouts and rafts to full-rigged carracks and caravels. At this point—we have reached the sixteenth century—I will break off in order to get to grips with the physics underlying sailing boat motion, to answer the following question: how do square-rigged sailing ships acquire power from the wind, and how close to the wind can they sail?

I will resume the tale of sailing evolution in chapter 3. My break point is in some ways arbitrary; I might have chosen the end of the nineteenth century, for example, because square-riggers were active up until then. Also, it is important for me to convey the continuity of evolution. There was no historical intermission in ship evolution during the sixteenth century; we will see how caravels evolved continuously into galleons, and from galleons into . . . well, the magnificent ships of chapter 3. Yet my decision to break at the Age of Exploration is not wholly random.

Most of the key ship-building ideas that led ultimately to the wonderful Yankee clippers and tea clippers at the very end of the sailing era were known in many of sixteenth-century Europe's seafaring nations. Ship evolution over the next 300 years was driven largely by other factors not connected directly with sailing. For example, larger cannons with longer ranges, commercial competition, and the perceived need for smaller crews all played a part, at one time or another, in shaping the ways in which sailing ships evolved between 1550 and 1870, as we will see. However, all the basic sailing ingredients were already in place: square and fore-and-aft (in the form of lateen) sails, stern rudders, carvel hulls, stays, shrouds, and lines. The period of evolution covered in the present chapter ushered in other technical developments: the magnetic (lodestone) compass, the astrolabe for estimating latitude, plus improvements in rope-making, water barrels, anchors, and naval charts. These small-step, behind-the-scene evolutions in the details of ship design and operation would continue to the present day.

So we pause at the birth of fully rigged ships to analyze how the square-rig sail plan extracts power from the wind.

Fly by night: *Unreliable, temporary.* A large sail that replaces several smaller sails and requires little attention. Used when sailing downwind at night.

Plumb the depths: *Sink very low.* Sailors used a plumb line (i.e., a rope with a lead weight at one end) to gauge depth in shallow waters.

Sound off: *Express loudly.* In shallow waters, a sailor measuring depth would periodically shout out the depth of the water in fathoms.

Taken aback: *Surprised, astounded.* A sudden change in wind direction could leave sails pressing against the mast of a square-rigger, impeded forward progress. The ship was said to be taken aback.

Analysis: Square-Rigged Ship Motion

Aloof: *Remote, superior in manner.* From the sixteenth century (probably derived from *luff,* "to turn the head of a ship toward the wind"): a boat of superior sailing capability that can stay upwind of other boats.

Bosun: *Boatswain.* From Old English *bat* or Old Norse *beit* (both meaning "boat") and Old Norse *swain* ("boy").

Doldrums: *Gloominess, stagnation.* From Old English *dol* ("dull"); equatorial latitudes where low winds often becalmed sailing ships.

Fathom: *A measure of six feet, or to gauge the depth of something—to figure it out.* From the Anglo-Saxon *faetm* ("embrace"), the length of a man's outstretched arms being about 6 ft; also German *faden* ("thread"). Measuring sea depth with an unmarked line, a sailor might haul in the line and count off 6 ft for each section pulled in.

I have pressed the pause button in my history of sailing ship evolution at a critical stage. By the sixteenth century, sailing vessels of the Atlantic tradition were on course to become the best that the world could produce,* as we will see in chapter 3. These ships were by and large square-rigged. History tells us that such vessels sailed effectively with the wind and were thus particularly well suited to traveling long distances across open ocean, pushed along by trade winds. But history also tells us that the later, more sophisticated square-riggers could also sail at least a

* But were not there yet—recall the Chinese junks of the early fifteenth century.

couple of points into the wind. This capability made them much more useful for getting about the globe and for maneuvering in combat. What is the physics behind the motion of square-rigged sailing ships? Can we understand quantitatively why they performed well and how close to the wind they could sail? These questions are the subject matter of the present chapter, in which I introduce you to the physics behind wind-powered sailing. As stated in the introduction, my emphasis will be on understanding the principles rather than on technical details. The ideas and physical principles established here will be built on and refined in chapter 4, much as designs for square-rigged ships were built on and refined after the Age of Exploration, as we will see.

Apparent Wind

The very first sailing physics topic that I need to convey to you is the idea of *apparent wind* velocity. This notion is absolutely fundamental for an understanding of sailing vessel movement. Fortunately, it is a simple idea and one that already will be very familiar to those of you who sail. I will state the case, nevertheless, because we will require the concept very soon when seeking to understand how wind provides drive for square-riggers. Even if you are thoroughly at home with apparent wind, there may be something for you to learn in this section, and it will serve to tell you about my notation.

Albert Einstein was a keen recreational sailor (fig. 2.1). This fact is pertinent to my discussion of apparent wind because apparent wind is all about relativity, a subject that the good professor (who appeared a little windblown most of the time) knew something about. Interestingly, the relativity theory pertinent to apparent wind is not that of Einstein but of a much earlier scientist whom Einstein greatly admired: Galileo. Galilean relativity is basic common sense: if you are driving along at 30 mph and another car is coming toward you at 40 mph, then your relative speed is 70 mph. Prof. Einstein would disagree,[1] but that's life; we will go with Galileo. Now, to extend this idea to a two-dimensional sea surface rather than a one-dimensional road, we need to describe velocities by *vectors*. The notion of apparent wind is explained simply in vector notation in figure 2.2a.

I need to be careful about nomenclature: to a physicist *speed* refers to how fast an object is moving, whereas *velocity* refers to both how fast

Figure 2.1. A weekend sailor on the water. Image courtesy of the Leo Baeck Institute, New York.

and in what direction it is moving. So, in figure 2.2a the arrows represent wind velocity, whereas the arrow lengths represent wind speed. The angle between boat velocity, $\underline{v}$, and true wind velocity, $\underline{w}$, is a_{vw}. (Throughout the book I will denote vector quantities like velocity with underlining.) The wind velocity felt by the moving boat is not $\underline{w}$ but rather $\underline{w}' = \underline{w} - \underline{v}$; the boat moves through the wind and so feels a wind velocity that depends on its own speed and direction, as well as that of the wind. Consider what happens if $a_{vw} = 0°$, in other words, if the boat is

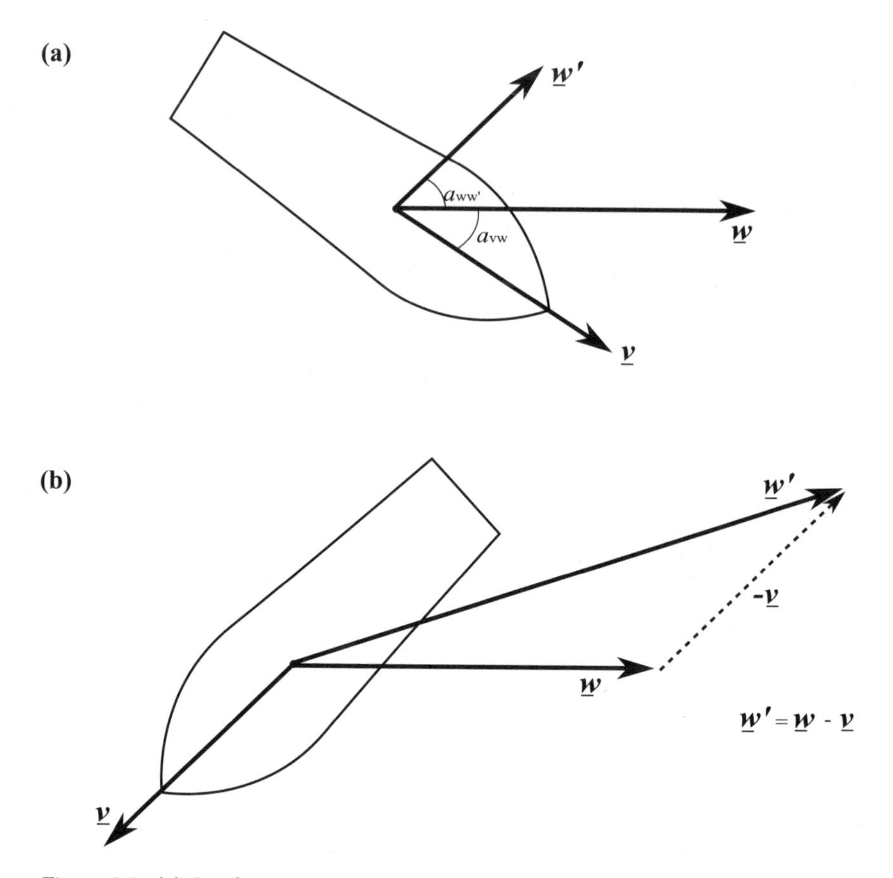

Figure 2.2. (a) Our boat moves with velocity $\underline{v}$, in a wind with true velocity $\underline{w}$. She feels an apparent wind $\underline{w}'$. (b) The boat has now changed direction and is heading upwind with the same speed: note how the apparent wind speed increases. (The dashed line shows how vector addition works.) In both cases the boat heading is closer to the apparent wind direction of origin than to the true wind direction of origin.

running before the wind. In this case she feels a following wind of reduced speed, as you might expect. If, as illustrated in figure 2.2a, the point of sail is a broad reach, then apparent wind velocity, $\underline{w}'$, has a different direction, as well as a different speed, from true wind velocity, $\underline{w}$. If, as in figure 2.2b, the boat is close hauling (i.e., sailing into the wind), the apparent wind speed is greater than true wind speed. In the two cases shown in figure 2.2 the boat speed is the same; only the direction is different. Yet this difference produces very different apparent wind velocities. The effect of boat velocity is always to make the apparent wind seem more head-on than the true wind, which is fairly obvious if you think about it.

Wind Force, Boat Force, and Isaac Newton

Now let me give our boat a name and a sail. She will be christened *Snoozing Goose,* and her single mast has been square-rigged. So she is a small ship with square sails that can be trimmed to best advantage, permitting movement upwind (to some extent) as well as downwind, as we will see. I will treat *Snoozing Goose*'s sail(s) as if they were a single flat sheet. Of course, this is not how real sails look, but it is a convenient approximation that permits us to readily specify sail orientation and makes the calculations that follow a lot easier. Figure 2.3a introduces the sail orientation angles relative to true wind (a_{sw}) and relative to apparent wind ($a_{sw'}$). I am also assuming that *Snoozing Goose* heads in the direction that she points—in other words, that there is no leeway, or sideways motion. This is a reasonable approximation, about which I will have more to say in a later chapter.*

In figure 2.3b you see that I have dispensed with the true wind—the boat senses only apparent wind—and have shown how wind force, $\underline{F}_{wind}$, converts into boat force, $\underline{F}_{boat}$. Wind force is just what the term implies: it is the force exerted by the wind on anything that gets in the way. Obviously, the direction of $\underline{F}_{wind}$ is the same as the wind direction. The magnitude of the wind force[2] is $F_{wind} = \rho A w'^2$, where ρ (Greek *rho*) is the density of air (about 1.3 kg m^{-3} or, if you prefer, 1.3 oz/ft^3), A is the effective area of the object (our boat) that is feeling the wind force, and w' is our old friend, the apparent wind speed. The wind force increases as the *square* of wind speed: double the wind speed and the force increases fourfold; triple the wind speed and the force increases by a factor of nine.

The area of sail that is struck by the wind is the maximum effective sail area A only if the sail and apparent wind point in the same direction (the angle $a_{sw'} = 0$, in the notation of fig. 2.3a). If the sail is set differently,

* Between figs. 2.2 and 2.3 there seem to be a lot of subscripts. Don't worry—I do not intend to involve you unnecessarily in arcane math notation, with this one exception (plus a few Greek letters), but subscripts will be used quite often because they remind us which forces and angles, etc., we are talking about.

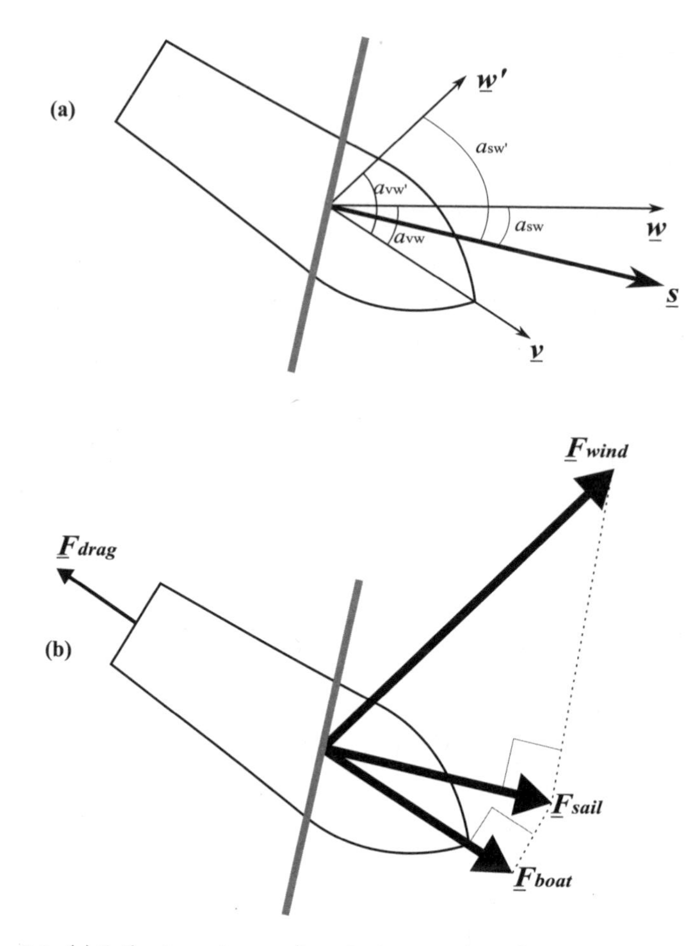

Figure 2.3. (a) Sail orientation angles relative to true and apparent wind. The sail of *Snoozing Goose* is here shown by the gray line: its setting is defined by the "normal" vector $\underline{s}$. The angle between sail and true wind is a_{sw}; between sail and apparent wind the angle is $a_{sw'}$. The angle between boat velocity and true (apparent) wind is a_{vw} ($a_{vw'}$). (b) Forces acting upon a boat. $\underline{F}_{wind}$ is the wind force; the projection of this force along the sail vector yields the force, $\underline{F}_{sail}$, felt by the sail. A second projection, of the sail force along the boat velocity direction, gives us $\underline{F}_{boat}$, the force pushing the boat forward. Hull movement caused by this force is opposed by hydrodynamic drag, $\underline{F}_{drag}$, as shown. (The component of $\underline{F}_{sail}$ that is perpendicular to $\underline{F}_{boat}$ is not shown, since it is here assumed to be cancelled by the effect of the keel.)

the apparent wind sees the projected area, which is less. So, for example, if the sail is at right angles to the wind (the sail is luffing), the wind sees practically nothing and so imparts zero force to the sail.

Now we have the sail force (the middle of the three forces shown in fig. 2.3b). We can see that it consists of two components. The boat force (often called *drive*), F_{boat}, is the component of sail force along the boat velocity direction; the lee force (not shown in fig. 2.3) is the component perpendicular to boat velocity. Here I will assume that the lee force is cancelled by *Snoozing Goose*'s keel and hull. Again, this is a simplifying approximation, but it is reasonable. One of the main functions of a keel is to reduce the leeway of a boat, and it achieves this because of its shape. The keel presents a broad area to the water if the boat is pushed sideways, but a much smaller area if pushed forward. So, put very simply, the keel helps to prevent the wind from pushing a boat sideways. I will have a lot more to say about the crucial role played by keels later on.

Thus, the boat force is a double projection of wind force: the component of wind force along the sail direction is projected along the boat velocity direction. This method of calculating wind force is called the *momentum flux* approach, or the *Newton's Third Law* approach. Already, a few readers will have spiraled into indignation bordering on apoplexy. They would claim (perhaps vehemently insist) that this method is wrong: we need the modern fluid dynamics concepts of vortices and circulation to properly account for the sail lift force that originates with wind force. Strictly speaking, these folk are correct. I will open up the can of worms labeled "aerodynamic lift" in the appendix. Those of you who are in a technical mood, or who are of a combative disposition, may wish to take a sneak preview of the appendix, but it is not necessary at this point. It will still be there when you reach the end of the last chapter. The simple approach that I am adopting here for square-rigged sails produces a pretty good approximation to the correct boat force. The analysis is a lot simpler and provides an understanding of how various parameters (such as wind speed, boat mass, and water resistance) influence boat movement. In chapter 4, I will compare this naïve momentum flux approach with a more modern style of calculation that treats sails as airfoils.

Momentum Flux and Effective Sail Area

I need to nail my colors to the mast before proceeding further and tell you a little more about momentum flux. The most naïve application of Newton's laws envisages the wind force on a sail as illustrated in figure 2.4a. Here the wind is viewed as a myriad collection of atoms and molecules, little ball bearings that collide and bounce off each other and off the sail. Between collisions they are considered to be quite independent of one another, and so move in straight lines. A sail in the wind encounters these ball bearings as shown: they bounce off one side of the sail, thus (when the effects of all the atoms colliding at all points of the sail are taken into account) exerting a force that is perpendicular to the sail. Were you to fashion a sail out of sheet metal and bombard it with a large number of ball bearings fired from a machine gun (I would recommend ear protectors), you would indeed find that the metal sail would experience the force shown. This force arises from ball bearing momentum that has been transferred to the sail. Note in figure 2.4a that the "spent" ball bearings have been deflected downwards.

This is too naïve. These days we do not regard air as a collection of little ball bearings, but instead see it as a fluid. Yes, the fluid consists of atoms and molecules, but they react with each other—each exerts a force on its neighbors—resulting in viscosity, compressibility, and other bulk characteristics of fluids that simply cannot be accounted for properly within the ball-bearing model. Considering the air to be a fluid, we find that the sail influences the wind flow as shown in figure 2.4b. The streamlines can be thought of as trajectories followed by constituent atoms and molecules of the air, but these lines are smooth—there are no sudden collisions. There is still a net downward flow toward the rear of the sail. The air pressure is increased on the windward side of the sail and reduced on the lee side. The mental picture that I am trying to convey here is of a fluid flow that is disturbed by the presence of the sail, causing the fluid to be deflected around the sail. This results in areas of increased and decreased pressure.

So, how do we calculate wind force imparted to the sail? In the naïve approach (fig. 2.4a) we calculate force as the product of air pressure and projected sail area. The actual area of the sail is $A = dh$, where d is the sail width (shown in fig. 2.4a) and h is the sail height (out of the page). The

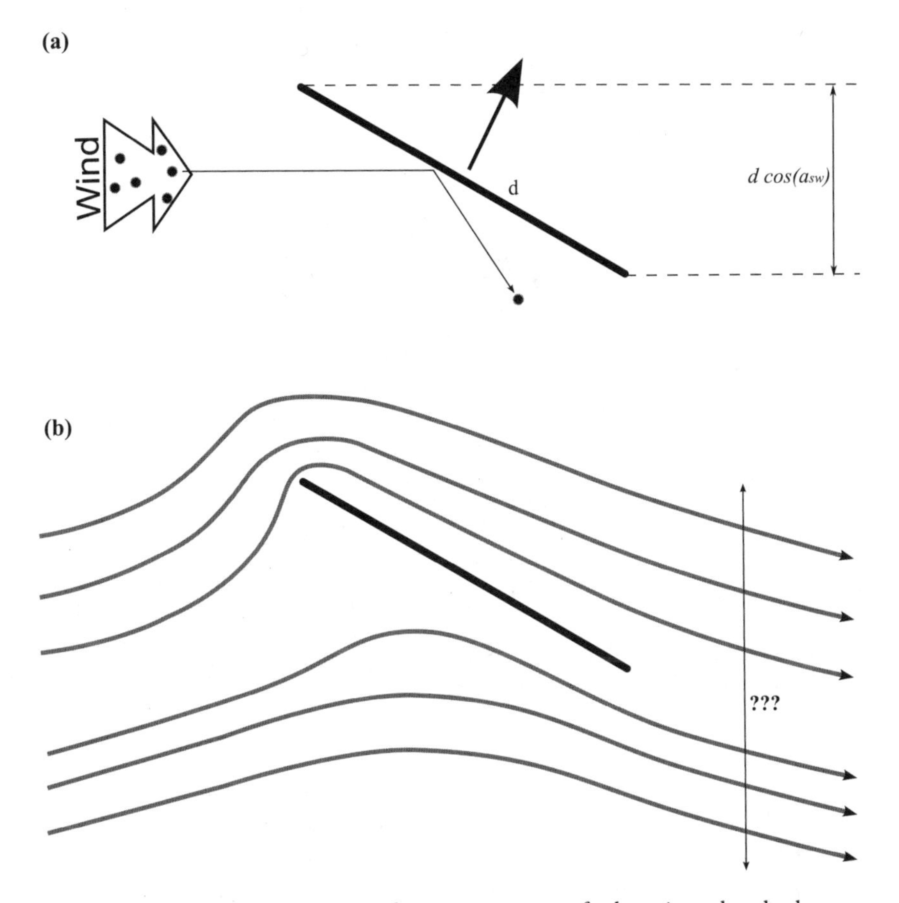

Figure 2.4. (a) The naïve version of momentum transfer has air molecules bombarding the windward side of the sail and so providing the sail with drive (arrow). (b) More realistically, air is regarded as a fluid that flows past the sail and is disturbed by it. Air momentum is shifted as a consequence. But how much air should we take into consideration?

projected sail area is then $dh \cos(a_{sw})$, where a_{sw} is the angle of the sail to the wind. In the modern momentum flux view, what is the sail area? You can see from figure 2.4b that the area of air disturbed by the sail extends beyond the physical projection of the sail area, so how much do we include? If height exceeds width, then the rather surprising answer is that *effective* sail area is $A = \pi h^2/4$. In other words, the area of air that we should use in our calculations depends on the height of the sail but not on its width or physical area. A justification of this result is provided in

Figure 2.5. The magnificent barque (or bark) *Europa* has a square-rigged foremast and mainmast. The masts are tall and the sail area is large, so she will run before the wind very well. Note the low aspect ratio (height to width) of the individual square sails: these will not be very effective beating upwind. Hence the fore-and-aft sails: the gaff-rigged mizzen and the triangular staysails. Thanks to Simona Manca for this image.

the literature cited in the bibliography (see especially Prandtl and Tietjens, and Waltham). Here we note the plausibility of it by considering the sail as an airplane wing and the wind force as the lift force exerted on the wing. It is well known that wings of the same geometrical area but different aspect ratios (length-to-width ratios) generate very different lift. This observation cannot be explained if the effective wing area is simply the geometrical area of the wing. Wings with a high aspect ratio (long, thin wings) generate the most lift.[3] Here, long wings equate to tall sails.

This brief exposure to (naïve) aerodynamics merely sounds the opening chords of the symphony; it is no more than an aside to give good reason for the notion of effective sail area. Square sails have a rather low aspect ratio compared with modern triangular sails of the same geometrical area (fig. 2.5).* Therefore, we can anticipate that modern sails generate more lift, which is why square sails were eventually superseded. To be continued.

Hydrodynamic Drag

So much for wind force. Before we can understand how the boat moves, however, we need to consider water resistance. I will gather together all the resistance forces that act on our boat and label them all "drag." Of these forces, the most significant is the resistance offered by water as the boat plows through it. There is also resistance due to aerodynamic drag on the sails and on the boat hull, but these are rather small by comparison. So, in the spirit of "enlightening approximations" that I promised you in the introduction, we will neglect all the resistance forces except for the hydrodynamic drag acting on the boat hull.

Hydrodynamic drag arises because the boat hull must push water aside in order to move forward. (It is harder to displace water than to displace air, which is why hydrodynamic drag is so much more significant than aerodynamic drag.) Another type of hydrodynamic drag arises from the action of surface waves, but in this chapter I will simply lump together all the resistance effects of water acting on our boat hull. Drag force depends on speed through the water as follows:[4] $F_{drag} = -mbv^2$. The minus sign indicates that drag opposes the boat force which drives our boat forward, as in figure 2.3b. The boat mass is m. The constant b is a drag factor and depends on hull shape and smoothness, and on water density and viscosity. (The parameter b is larger for molasses than for water—a fact of limited value in the world of sailing.) The speed, v, in our drag equation is that of the boat through water. Putting together the force of the wind acting on the sails and the force of the water acting on the hull, we obtain the total force acting on our boat:

$$F = F_{boat} - mbv^2. \tag{2.1}$$

*In fact, the aspect ratio is not very important when running with the wind, which is what square-riggers were designed to do. Consequently, the sails did not run the full height of a mast but were split up, as shown in fig. 2.5. This was also much more practical. High aspect ratio matters when reaching or sailing to windward, when sails present a leading edge to the wind. Roughly speaking, more edge means more lift force.

A physicist would say that he has solved the problem once he gets to this point: the rest is just math. The total force equation is converted into an "equation of motion" for the boat by applying Newton's laws; this results in a differential equation which in general is difficult to solve.[5] However, we can draw a number of useful conclusions from this differential equation without having to solve it. The rest of this chapter is devoted to showing these results (skipping the mathematical derivations).

Equilibrium Speed

The first and most significant conclusion that we can draw from this analysis concerns the maximum speed that the wind force can impart to our boat. Here we are ignoring the water wave phenomenon that gives rise to hull speed, discussed in a later chapter, and concentrating simply on the speed that *Snoozing Goose* can extract from the wind. By setting her sails appropriately, she can catch the wind and make forward progress in any direction except straight upwind. The math analysis of equation (2.1) leads immediately to the following conclusions:

1. If $a_{sw'} = 90°$, so that the plane of the sail is parallel to the wind direction, then the sail provides no wind power. Our boat is *in irons*, an embarrassing situation characterized by the sail luffing (flapping aimlessly) while the boat drifts.
2. For a more sensible sail setting, from a standing start the initial force imparted to the boat by the wind is $F_{boat} = F_{wind} \cos(a_{sw'})\cos(a_{vw} - a_{sw'})$, which reaches a maximum value for $a_{sw'} = a_{vw}/2$. This observation tells us that there is a best choice for setting the sail, to coax speed out of the wind. This setting may change as the boat speeds up because the apparent wind direction changes with speed.
3. If *Snoozing Goose* is running before the wind—i.e., heading straight downwind—she will pick up speed until she reaches an equilibrium speed limit $v_{eq} = w/(1+\sqrt{\beta})$, where $\beta = b/\rho A$. Once she reaches this speed, the wind force drops to zero and she cannot accelerate, so v_{eq} is her maximum downwind speed.

Conclusion 3 is the most important from the perspective of determining sail performance, though from the practical point of view conclusions 1 and 2 need more immediate attention from the helmsman, who must ceaselessly monitor his heading and the apparent wind direction to maintain or increase his speed through the water. The parameter β (Greek *beta*) is a dimensionless combination of other parameters that will crop up again in this book. If there is no hydrodynamic drag ($b = 0$ and so $\beta = 0$), the equilibrium speed becomes equal to the true wind speed. This observation accords with experience and common sense: a sailboat running with the wind cannot move faster than the wind and attains wind speed only if there is zero drag. For realistic situations where b (and so β) exceeds zero, the maximum downwind speed is less than wind speed. In principle *Snoozing Goose* can move faster than the wind for a broad-reach point of sail, heading downwind at, say, 45° from the apparent wind direction, but for most boats the drag factor is too large for this to occur. In figure 2.6, I plot the equilibrium speed for other points of sail, calculated from equation (2.1), assuming realistic values for drag factor. *Snoozing Goose* can move at about one-quarter of the wind speed, at best, when running downwind. The calculations confirm what history told us in chapter 1 about square-riggers: they are at their best when bearing away (fig. 2.7). She can travel across the wind (beam reach) and diagonally downwind (broad reach) fairly well (fig. 2.8), but her speed drops when heading up. Sailing into the wind is possible, but her best speed is only about 10% of wind speed when she is on a close reach, at about 45° or 50° from upwind. Closer to the wind her speed drops away to nothing.

Conclusion 3 also tells us how equilibrium speed depends on boat mass and effective sail area. Unsurprisingly, we find that heavier boats move more slowly, while those with more sail travel faster. Equally unsurprisingly, more wind means more speed. This kind of commonsense reality check is usual when physicists test their theoretical predictions in the real world: the obvious, sensible results have to be predicted correctly. Correct prediction of the obvious gives us confidence that the less obvious results are reasonable.

Bear in mind that all the physics I have applied so far (the momentum flux analysis) has gone into deriving equation (2.1); everything else that follows from this equation is just math. Think of the math as an orange juicer: it extracts juice but doesn't have anything to do with how much juice is in the orange to begin with. So, mathematical analysis helps us to squeeze a little more juice out of equation (2.1); it aids us in seeing the scientific content of equation (2.1), but does not add to it. Quite often

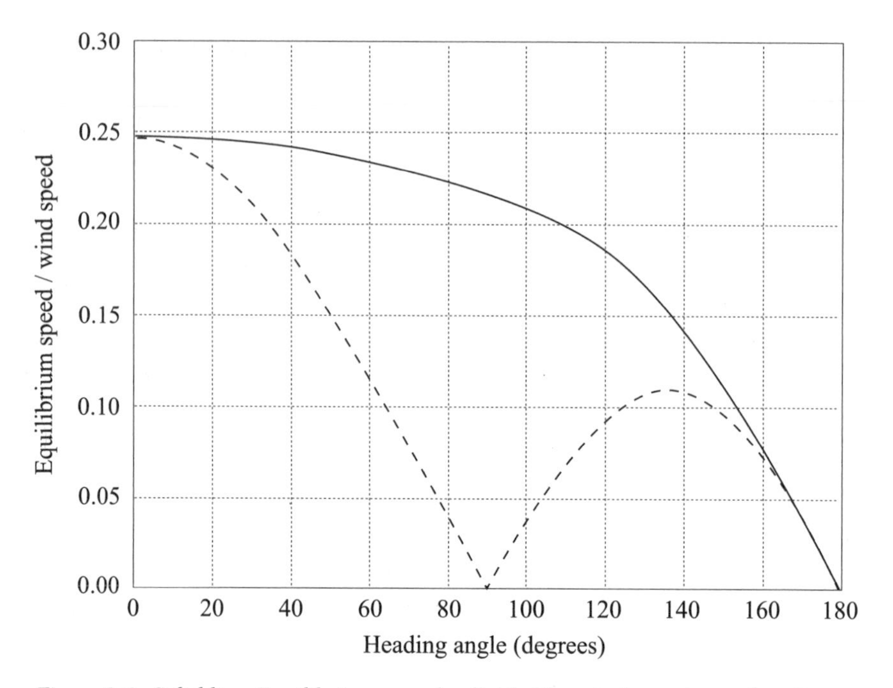

Figure 2.6. Solid line: Equilibrium speed v divided by wind speed w vs. heading angle a_{vw} for typical square-rig parameters. When running downwind ($a_{vw} = 0°$) our boat can move at about a quarter of wind speed. She travels across-wind ($a_{vw} = 90°$) quite well, but heading directly into the wind her speed drops away to nothing. *Dashed line:* Upwind and downwind speed. This is the component of boat speed along the wind direction. Across-wind ($\theta = 90°$), of course, our boat makes zero progress upwind or downwind by definition. The rate at which she makes progress upwind peaks (at about 11% of wind speed) when her heading is about 135°, i.e., 45° or 4 points off the wind.

this basic fact is overlooked; even some professional scientists mistake mathematical analysis for scientific reasoning. Mathematics is the language, not the statement.

What else can we explain by analyzing the equation for square-rigged boats via the momentum flux approach? First, I can demonstrate to you the old adage about sailboats "creating their own wind." If the boat is heading to windward, as it picks up speed, the apparent wind speed increases; this, in turn, increases the force that the wind applies to the sails, which makes the boat go faster—up to a point. The math tells us that the initial acceleration of the boat (i.e., acceleration from a standing start) increases faster and faster in certain cases, as shown in figure 2.9.[6]

Figure 2.7. This tall ship (the *Christian Radich*) has square-rigged mainsails and triangular staysails. She is running before the wind, which is what square-riggers do best. Image from Wikipedia.

This increasing acceleration occurs only when steering to windward, and then only for certain sail settings. In most instances the initial acceleration slows down as the speed increases (for example, we have already seen that it slows to zero as the equilibrium speed is approached). Figure 2.9 applies only initially; as our boat speeds up, the area of "creating wind" in figure 2.9 shrinks and then disappears, at which point the accel-

Figure 2.8. These square-riggers are sailing on a broad reach or a beam reach and can take such points of sail in their stride. Thanks to Darillo for providing these images.

eration slows and speed levels off. As also shown in figure 2.9, the best initial choice for sail orientation ($a_{sw'} = a_{vw}/2$, determined earlier) does not overlap with the region where square sails "create their own wind." I guess that square-riggers were just not meant for sailing to windward.

Second, you may have wondered how quickly *Snoozing Goose* approaches her equilibrium speed. This question is of practical importance because a boat that accelerates only very slowly may take hours to reach that speed, whereas a zippier vessel may take only minutes and so will

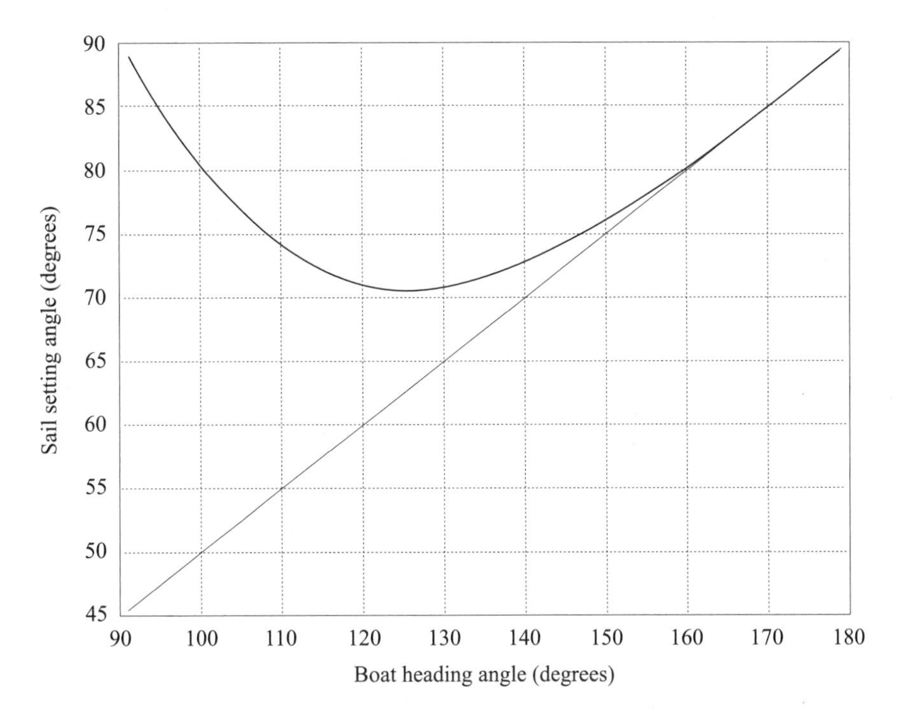

Figure 2.9. The initial acceleration of *Snoozing Goose* increases from the starting value if the sail setting angle (vertical axis, $a_{sw'}$) and boat heading angle (horizontal axis, a_{vw}) lie above the bold line. In this case, the *Goose* is creating her own wind. For ($a_{sw'}$, a_{vw}) values that lie below the line, the acceleration slows down. Unfortunately, the best choice for initial sail setting angle, $a_{sw'} = a_{vw}/2$ (thin line), is below the line, and so square-riggers do not easily create their own wind when close-hauling.

maintain a faster average speed. Well, the answer to this question requires number crunching, which is to say putting equation (2.1) into a computer and churning out an answer, because the equation is generally solvable no other way. However, such an approach is against the spirit that I have adopted for this book. I don't want to turn these pages into an arid technical treatise; if a simple answer cannot be given, I will leave the question aside, as enlightenment is unlikely to be swift amid a morass of data. However, for the particular case of a boat running straight downwind, the equation can be solved without computers,[7] and out pops the following characteristic time:

$$\tau = \frac{1}{2w}\sqrt{\frac{m}{\rho A b}}\,. \tag{2.2}$$

The longer the characteristic time, the slower our boat approaches its equilibrium speed. Conversely, short characteristic time means a faster approach to equilibrium speed. So we see that, unsurprisingly, a bigger sail area, A, means a more rapid approach to equilibrium speed, and a bigger boat mass, m, means a slower approach.

More unexpected, perhaps, is the dependence on apparent wind speed and drag factor: our boat approaches her maximum speed faster in a strong wind than in a gentle breeze and also approaches faster as the hydrodynamic drag increases. We can explain the drag part readily enough. A large drag factor means low equilibrium speed, as we saw in conclusion 3, so it doesn't take long to get there from a standing start. The wind speed dependence is more of a challenge. Recall that we are running before the wind and that equilibrium speed increases with wind speed. If the wind speed is 10 knots (about 5 ms^{-1}), *Snoozing Goose*'s equilibrium speed is about 3 knots (assuming a realistic value of $\beta = 5$), whereas in a 20-knot wind her equilibrium speed is 6 knots. Equation (2.2) is telling us that it takes *half the time* to get within spitting distance (say 99%) of 6 knots in a 20-knot wind than it takes to get to 3 knots in a 10-knot wind! How can that be so? The answer has to do with the energy possessed by the wind, and the power that it can transfer to our boat through her sails. A boat that is initially standing still will be propelled downwind with power that is proportional to the square of wind speed.[8] The 20-knot wind transfers four times as much power to our sails as the 10-knot wind. This is why it accelerates the boat more quickly.

Beating to Weather

A ship or boat is beating to windward or beating to weather if she is moving upwind—*beating* because she is heading into the waves, and her hull takes a pounding. We have seen in figure 2.6 that a square-rigger can indeed make progress against the wind, but slowly. Suppose we want to travel due west in a westerly wind.* Figure 2.6 and common sense both tell us that we cannot simply steer *Snoozing Goose* westward, because the wind would drag her eastwards. She can, however, sail northwest or southwest. With sails close-hauled she might be able to get within 45°

*A westerly wind comes *from* the west, but an onshore breeze blows *towards* the shore.

Figure 2.10. The *Mir*, a fully rigged Russian sail-training ship that can close-haul to within about 37° of the wind. No doubt the fore-and-aft staysails help, but the large bracing angle of her yards is crucial to the *Mir*'s capabilities to windward. I am grateful to Simona Manca for providing this image.

of the wind;* the square-rigger shown in figs. 2.10 and 2.11 regularly achieves this, though we will see that—and why—the fore-and-aft boats do better upwind. So she can head west by zigzagging back and forth, as shown in figure 2.12. Each time she changes direction (or changes tack, in a maneuver known as *tacking* when heading upwind), her sails must (quickly) be reset. More on tacking, shortly. First, I want to see what our momentum flux method of calculation can tell us about the optimum tacking angle.

We have seen (conclusion 2) that a good choice for sail direction is $a_{sw'} = a_{vw}/2$. Recalling the way that wind force is projected to provide drive (fig. 2.3b), you can see that sails which are set as in figure 2.12 will indeed drive the boat forward against the wind. She is close-hauled. But what value of a_{vw} is the best choice? How close to the wind should we sail? The best a_{vw} is that value which maximizes the speed upwind, i.e.,

*I.e. four points off the wind.

Figure 2.11. Mast and yards of the *Mir*. Thanks to Simona Manca for providing this image.

maximizes $|v_{eq}\cos(a_{vw})|$, and this corresponds to $a_{vw} = 132°$, which means steering 48° from the wind.[9] Here I have assumed that our boat is sailing at her equilibrium speed. She loses speed when changing tack, as we will see, and so my assumption implies that the length of each straight-line section of figure 2.12 is sufficient for us to get back up to equilibrium speed and spend most of our time there. This result—a heading angle of

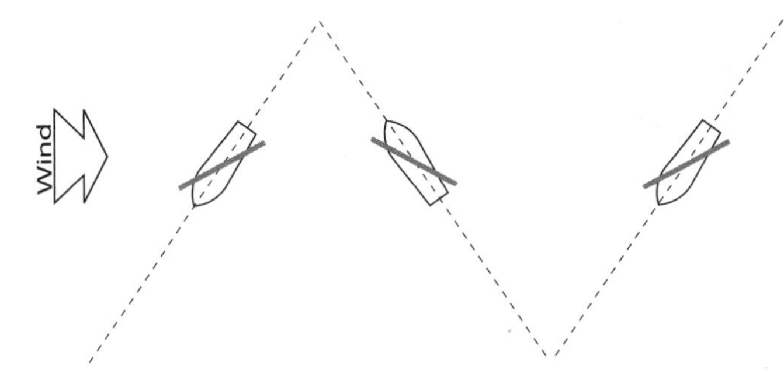

Figure 2.12. Here *Snoozing Goose* switches from starboard tack to port tack and back to starboard tack as she beats upwind. Tacking involves switching her sails from one side to the other at each turn.

48°—should be regarded as approximate, given the approximate nature of my calculations. So, let us say that the best heading angle is somewhere in the region of 45°–50° from the wind. This looks very plausible.

There are two methods of changing tack: tacking and jibing. These methods, illustrated in figure 2.13, highlight a difference between the old square-rigged ships such as *Snoozing Goose* and the modern Bermuda-rigged vessels that will be analyzed in chapter 4. In figure 2.13a *Snoozing Goose* switches from a port tack to a starboard tack by heading through the wind (i.e., coming about). Her bow, for a brief moment, is facing directly upwind. During this interval, it is important that her sails are slackened or reduced, so that they luff—i.e., offer minimum resistance to the wind. This is because they are not helping but hindering forward progress at the instant *Snoozing Goose* faces the wind. Momentum alone carries her forward. The sails are useless, and only the rudder is helping her bear away from the wind, to a starboard tack. If she takes too long passing through the wind, all momentum is lost to drag, and the rudder (which requires speed to be effective) becomes useless: *Snoozing Goose* is in irons. Assuming this does not happen, and that she has enough momentum to turn her bow through the wind, she sets sail for the starboard tack, and the wind again fills the sail and drives her forward.

Coming about is a maneuver that requires precise timing and, especially for a large ship such as a fully rigged ship of the line in the Age of Sail, requires many hands working frantically to set sails appropriately at the correct instant (fig. 2.14). Square-riggers cannot sail very close to the

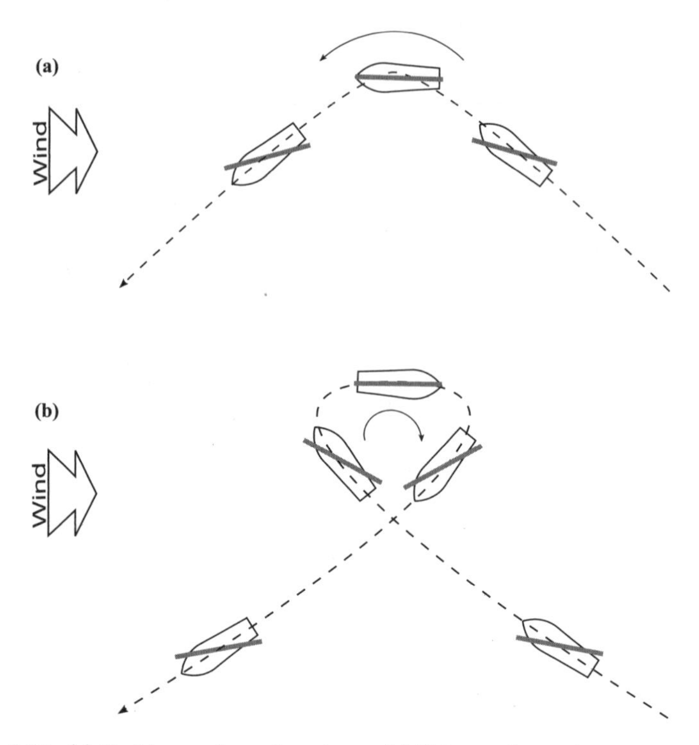

Figure 2.13. (a) Tacking and coming about; (b) jibing and wearing.

wind, and so they must rely on momentum alone for a large part of the turning maneuver—a distinct problem for slow-moving vessels facing rough open seas. For this reason, square-riggers prefer to jibe.

Jibing is illustrated in figure 2.13b. You can see that more open water is needed for this maneuver. The stern, not the bow, of the ship passes through the wind (in the case of square-riggers, this maneuver is called *wearing about*). Sails are always in use—no danger of getting into irons. As the illustration makes clear, precise sail-handling is still required, though in practice, wearing requires fewer hands than does coming about. So, square-riggers (especially large ships in open seas) change tack by jibing.

Small, swift modern yachts can sail close-hauled to within about 30° of the wind (and high-performance racing vessels can sail closer), as we will see, and less effort is required to shift the sails, though rather careful timing is still called for. So, modern fore-and-aft boats can tack or jibe, as circumstances dictate.

Figure 2.14. The complex rigging of square-rigged sails. Many hands are needed to reset these sails, and this is one reason why sailors prefer to jibe rather than tack these vessels. Thanks to Darillo for this image.

My description of changing course is incomplete, especially for larger ships with more than one mast, because another factor—torque—comes into play. Torque is the subject of chapter 5, and so I will defer a detailed explanation until then.

Weasel Words

My analysis shows how wind power drives a ship via the simple mechanism of momentum flux. Calculations required by the momentum flux approach are greatly simplified compared with those required for the full aerodynamic theory. (Later I will delve a little deeper into the modern approach concerning lift and drag.) Despite the simplifications, we have seen how a square-rigger can sail with, across, and against the wind. Momentum flux is readily conceptualized and accounts in a rough sort of way for much of the sailing behavior of square-rigged ships that is observed on the water.[10]

Quite apart from the theoretical simplifications, however, I have made some more practical simplifications to render the analysis more

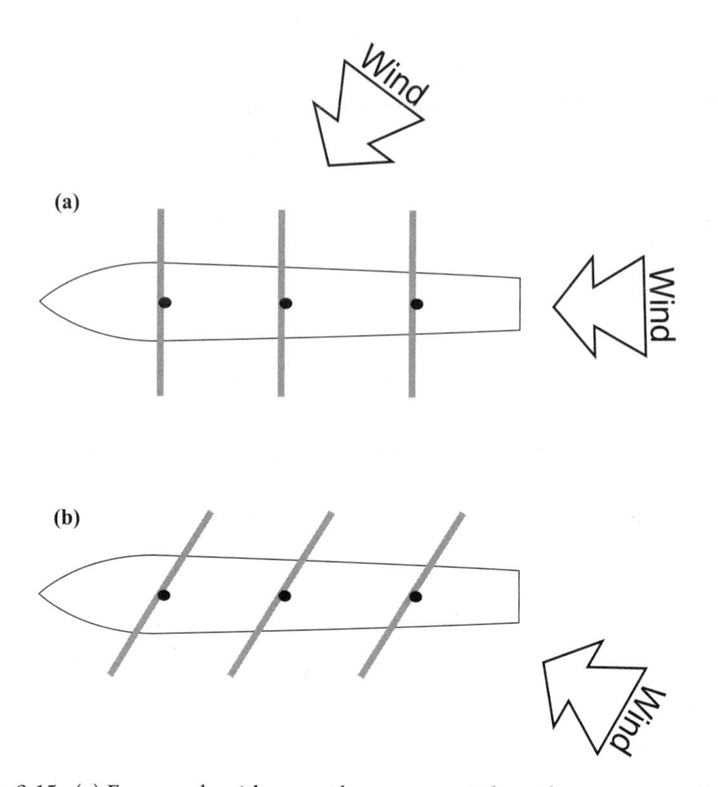

Figure 2.15. (a) For vessels with more than one mast the sail area presented to the wind depends upon wind direction in a complicated way (even with our simple momentum transfer approach) because windward sails screen or partially screen the leeward sails. (b) The sail area depends not only upon wind direction but also, of course, upon sail orientation. For these reasons, my analysis was restricted to single-mast ships, for which the projected sail area is easily calculated.

presentable. In this section I will discuss these simplifying assumptions. Some of them, no doubt, will have already occurred to you. First, the analysis assumes that *Snoozing Goose* has only one mast. If she had two or more masts, the analysis becomes a little more complicated. Thus, if the sails of one mast block or partially block the wind, preventing it from filling the sails of the other masts, the equation we used for effective sail area needs to be changed. From figure 2.15 you can see that the area of canvas that is presented to the wind depends on wind direction in a more complicated way than the simple projected area $dh\cos(a_{sw})$ that I assumed earlier. The manner in which sail area depends on sail orientation to the apparent wind now becomes different for different ships. The length of each yard (i.e., the width of the sails) and the spacing between masts become relevant. My analysis avoided these complications for ease of presentation, though I could just as easily have included them.

More difficult to handle is the changing shape of the sail due to wind pressure. A sail is not a rigid airfoil: it fills out or luffs; it changes shape with changing wind direction and speed. As suggested by figure 2.16, the wind force distribution across a sail is not uniform, and the effect of wind is not just to drive the sail—and the ship—forward, but also to twist the sail. This twisting force, or torque, would cause the yard to swerve around the mast, but for the bowlines. These lines, attached to the leeches, prevent twisting and keep the sail from shaking, especially when close-hauled. Without bowlines, it would be well nigh impossible for a square-rigged ship to head any closer to the wind than a broad reach. The bowlines transfer torque to the hull and so in general lead to a change of direction unless countered by the rudder. These effects have been ignored in my analysis.

Two or more sails can have other effects than changing the projected area. The so-called slot effect can come into play when the yards are turned to a large angle, so that they are running more fore-aft than across the hull. In such a configuration, square-rigged sails behave more like fore-and-aft sails, and the slot effect belongs firmly in the fore-and-aft analysis. We will return to this phenomenon in chapter 4. Here I simply note that the wind flowing past a sail can be significantly influenced by the sail, so that when it then impinges on a second sail the apparent wind direction has been altered. This slot effect is beneficial and is not accounted for in my momentum flux analysis.

It seems that historical square-riggers did not achieve their maximum potential for sailing to windward. I indicated earlier that in ancient times they might manage sailing one point into the wind (seven points away from the wind), and that by the Age of Exploration or the Age of Sail they could get within six points of the wind. But our analysis, and the performance of modern square-rigged vessels such as *Mir* (fig. 2.10), show that four points is possible—i.e., a square-rigged vessel should be able to sail within 45° of the true wind. Why the shortfall? Clearly the historical vessels were less efficient than modern ships: they would have made more leeway and suffered from greater hydrodynamic drag. Also, they might not have been able to angle their yards so much. The orientation of the yard to the hull is measured by the brace angle, which is zero

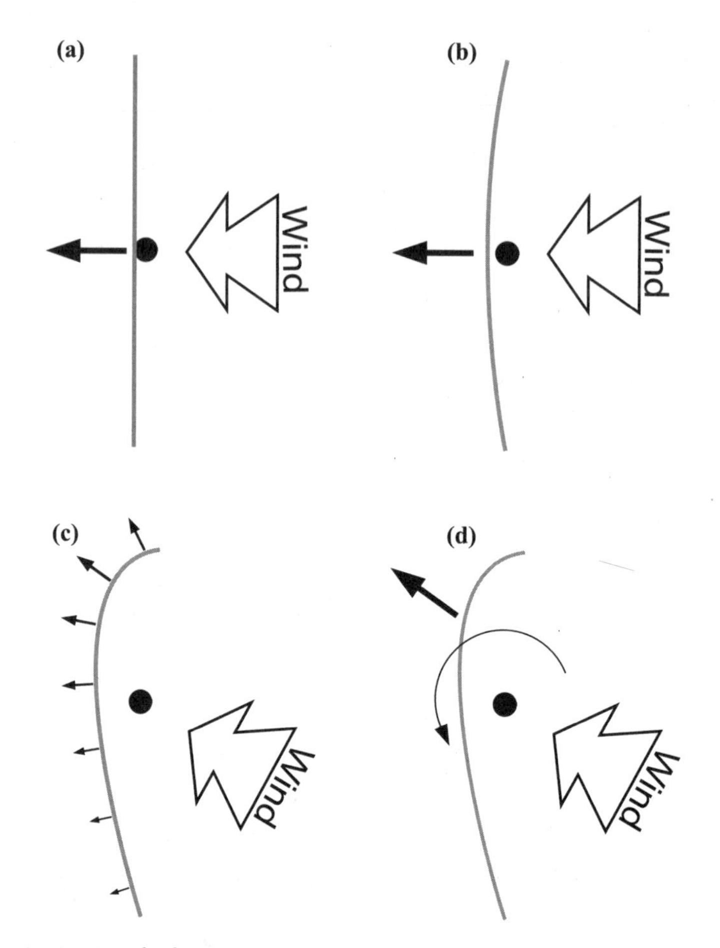

Figure 2.16. My calculations employed the simplifying assumption (a) that sails are flat. In the real world they change shape with the wind. This is unimportant when running with the wind (b), but for all other points of sail the sail shape is asymmetrical, resulting (c) in a markedly uneven distribution of wind force across the sail. A theorem in Newtonian physics says that the result of any number of such forces is equivalent to a single force and a torque, as in (d). So, we are allowed to represent the sail force by a single vector; however, sail shape can induce unwanted torque as well as the desired drive.

when the yard is across the hull, perpendicular to the longitudinal axis of the hull. The *Mir* can achieve brace angles of 60° on either side, which goes a long way toward explaining her performance, but historical vessels could not get close to this figure. For many ships the mast shrouds would have gotten in the way at large brace angles. In the 1840s the

British Royal Navy experimented with sailing ships to see if they could improve on their windward performance, and one very successful experimental vessel achieved brace angles of 70°. No doubt, the lessons learned from the trials of such experimental ships would have been applied in time, but the advent of steam took the wind out of their sails, so to speak.

There are other effects that I have ignored (for example, wind shear due to friction at the water surface), but the point is made, I think. It is important to understand that judicious simplification yields significant benefits: we obtain a good approximation to the truth behind a complex process without horrendous number crunching. This approximation, in turn, enables us to gain insight into the physical principles that might otherwise be obscured by reams of computer printouts. Here, we have learned that momentum flux is behind much of the observed behavior of square-rigged ship motion. Injudicious simplification can be misleading: had our momentum-transfer approach led to predictions not borne out by observation, we would have to conclude that this approach is invalid. Practical factors such as the slot effect and sail profiles do indeed influence ship performance, but as a first approximation can be ignored because they are not the main feature. In physics, progress often depends on knowing which approximations can be made. Without them the universe would be far too complicated for us humans, even weekend sailors like Einstein, to comprehend.

Heave To

At this point, we will bring *Snoozing Goose* to port. She has served her purpose of illustrating how the broad features of square-rigged ship movement can be understood in terms of simple physical principles, from the perspective of momentum flux analysis. The next chapter tells the story of *Snoozing Goose*'s successors, leading up to modern fore-and-aft rigged yachts. The design of these vessels reflects a better understanding of the underlying physics, and so our analysis will evolve in parallel.

Cut and run: *Make a hasty retreat.* Cut the lashings that held sails to yardarms. The sails would unfurl quickly and draw wind, thus enabling escape from an enemy ship fast-approaching or one identified as an enemy ship only when very close by.

In the offing: *At hand; likely to happen.* Since the sixteenth century, "offing" has meant the area of sea visible from shore. So a ship that appears in the offing is about to make port.

Rummage sale: *Sale of secondhand goods.* From the sixteenth century: "rummage" meant unloading cargo in port so that it could be stowed more effectively. The term derived from Old French "arrimage," meaning the arrangement of cargo in a ship. Damaged freight was sold at discount in arrimage sales. By the nineteenth century "rummage" had come to mean the formal inspection of cargo by customs, so that a "rummage sale" referred to confiscated or unclaimed goods.

Three sheets to the wind: *Very drunk.* A sheet is a line used to trim a sail. If the sheet was damaged, it blew in the wind and the sail billowed. A ship with three sheets in the wind would look bedraggled and would stagger uncontrollably.

3

Evolution: From the Age of Sail to the Modern Yacht

Chock-a-block: *Filled to capacity, packed.* In the age of sail, pulley blocks were used in the rigging of ships. During close-hauling, blocks were "choke-a-block"—they could be brought no closer together.
Head: *Toilet.* From the eighteenth century. The toilet onboard a ship is referred to as a head because it was located at the front of the ship, by the bowsprit, where splashing of the waves would naturally clean the area.
Scuttlebutt: *Rumor.* U.S. Navy slang based on a water fountain or barrel around which sailors would gossip. To "scuttle" is to sink a ship by holing the hull; "butt" means cask. In the days of sailing ships, water barrels were holed to provide access; hence, they were called scuttlebutts.
Skyscraper: *A very tall building.* Originally a triangular sail at the very top of a mast, set to catch a light wind.

Sail Plans

The first school of oceanic navigation was established in Portugal in the early fifteenth century.* Clearly this action was part and parcel of a new view of the world: with sharp eyes fixed on the horizon, the Portuguese sent out their little caravels to write the first chapter of the Age of Exploration. Madeira, the Azores, and the New World; rounding the Cape of Good Hope and across the open ocean to India; Southeast Asia,

*This chapter covers the period from about 1550 to 1880, which spills beyond the period usually considered to be the Age of Sail by about a generation on either side.

China, and Japan; circumnavigation. The organization that was required for these voyages, and for the maritime navigation school, heralded the future for sailing ship evolution. From these times, because of a multitude of converging though disparate circumstances, ships grew larger. Increased size meant increased canvas area, more complex rigging, and more crew—hence, more planning and organization. Increased size also meant more detailed ship design. Large ships must be planned from the drawing board; they cannot be the product of a single craftsman or of a small group, all of whom are familiar with every aspect of the vessels they build. Thus, increased ship size entailed increased planning and organization. The late sixteenth century gave rise to full-time professionals who today would be called naval architects. The signs were on the wall (actually, the bulkheads) much earlier: carracks needed to be constructed with a more complex and complete skeleton, with planking framed on the ribs all the way from gunwales to keel, rather than the earlier and simpler carvel framing.

The decades around 1600 CE saw the first examples of standardization of ship design, with codified rules and manuals for construction, for example, laying down the dimensions of ship components.* It is easy to see why such codification and standardization was becoming necessary, yet national differences still persisted. For example, the French pursued a "rule of measure" design, whereas the English had their own "rule of thumb" design. At this stage the French approach was better, and English galleons often copied French galleon designs (as earlier they had copied carrack designs).

The circumstances that drove the shipbuilding nations of Europe to larger and larger vessels were many and varied, usually with no direct connection with ship architecture. Instead, a complex interaction between cause and effect resulted, spurring both. For example, one reason for the increase in ship size, and in particular for the replacement of the popular though clumsy carrack by the galleon, was the improvement in artillery. Cannon had been placed on warships and on some merchant-

*Thus, Henry Bond in 1664 published a very detailed book entitled *A Plain and Easie Rule to Rigge any Ship by the length of his Masts, and Yards. Without any Further Trouble.* One small example: "For the spritsail yard: halyards—three times the length of the yard; lifts—three times the length of the yard; clulines—two times the length of the yard; buntlines—two times the length of the yard; sheets—three times the length of the yard; pennants—one-third of the yard."

men since the earliest days of artillery in Europe. (The *Mary Rose*—the flagship of Henry VIII, you may recall—possessed cannon as well as archers.) From these early beginnings, cannon had grown in size as well as accuracy and effective range. These developments impinged on naval design and tactics. I will get to the tactics later. As for design, it is clear that larger cannon required larger ships. An arms race ensued: more cannon to defeat enemy vessels required a larger ship, and the enemy responded in the same vein, so that bigger cannon were needed to sink his ships. Another major spur to the increase in ship size was international trade. Exploration led to more trade over longer distances. This in turn required more seaworthy vessels carrying larger amounts of cargo. A 300-ton ship in 1400 CE was considered to be large; by 1425 the same description was applied to ships of 720 tons. Dynastic and religious wars that characterized the Middle Ages gave way, by the early seventeenth century, to trade wars—beginning with the three Dutch-English wars between two major maritime powers. War over trade entailed armed merchantmen and convoys guarded by warships (figs. 3.1–3.2). Henceforth the major trading nations became the dominant ship designers and developers.

The changes stimulated further development. Warship and merchant ship design parted company during this period, and ships became specialized to carry more cannon or more cargo. As we saw in chapter 1, warships in European antiquity were distinct because they were rowed, not sailed; that distinction had blurred with the demise of galleys following the collapse of the Roman Empire. Now it reasserted itself. A sign of the changing times came in 1552 when Spain prohibited ships of less than 100 tons from traveling to the New World. This political proscription may have been aimed at restricting the diminutive Portuguese caravels, but it indicates the way that ships were growing. (At that time Spain was heavily into galleons, as we will see.) In 1587 the minimum requirement was increased to 300 tons.

What of the sails themselves? Not only was the planning of ships and sailing growing and changing during and after the Age of Exploration, but the ships' sail plans were evolving in parallel. Over the entire history of sailing ships, the development of sails themselves occurred mostly as a series of small technical innovations. Thus, square sails in antiquity consisted of, say, woven flax fiber suspended from a single yard (the top spar), whereas square sails in the nineteenth century hung between yards

Figure 3.1. Reconstruction of the large Dutch merchant ship *Batavia,* which sank in 1628 off the coast of Australia on her maiden voyage following a mutiny. Merchantmen of this period were much larger than their predecessors and traveled much further. *Batavia* was bound from the Netherlands to the Dutch East Indies. She displaced 1200 tons, was 187 ft long, and had a length-to-beam ratio of greater than 5:1. She carried 41 crew and passengers, and 24 cannon. Thanks to Mike Cawood for this image.

with a complex pivoting arrangement to increase the bracing angle and so permit sailing to windward. The running rigging evolved in complexity to cope with the increased capabilities of the sails, and sailcloth evolved from hemp or jute to more closely woven cotton.* Sail-makers

*Cotton sails became popular in Europe after 1851, the year in which the U.S. racing yacht *America* decisively defeated a fleet of British yachts. *America* had cotton sails whereas the British yachts did not; at the time this difference was reckoned to be a significant factor in the result of the races. Cotton can be woven more closely than hemp or jute, so it does not lose wind through the pores and does not stretch out of shape so easily. On the other hand, cotton sails are stiffer and more difficult to handle.

Figure 3.2. A detail of the *Batavia*'s hull that gives an idea of the sturdiness of these seventeenth-century armed merchant ships. This replica sailed from the Netherlands to Australia in 1999. Thanks to Mike Cawood for this image.

learned to cut the sails to shape and sew them together in the optimum shape (today we would say "the most aerodynamic form") to take advantage of gentle breezes as well as strong winds. Primary sail types (those used to drive a ship forward) proliferated* and, grouped in combination on multimasted ships, enabled a ship's sails to be trimmed to suit conditions: maneuvering in calm summer weather in a sheltered harbor, open ocean voyages in a winter storm, racing in competition, or maneuvering in battle. The increasing sophistication of sail disposition and use over the centuries was of course matched by a similar increase in rigging development, though much of this history of complex and interdependent components is obscure.

*We have already encountered square, lateen, and sprit sails. Later developments include gaff, lug, and of course the modern Bermuda sail—a total of six primary classes.

Galleons

I have already mentioned one reason for the emergence of large, well-remembered galleons: they housed large sixteenth-century cannons better than did the earlier carracks. More generally, the galleon was a response to the inefficiency of the carracks. The need for bigger ships also relegated the nippy little caravels to history: enlarged caravels did not sail nearly so well. The galleon, a Spanish innovation that dominated European waters for two centuries, solved the problems of the times: it carried a lot of guns and cargo, and it sailed well.

The word *galleon* (from the Italian *gallioni,* meaning a galley with superior sailing qualities) is a generic label for a wide range of large ships built to serve different needs. The original fifteenth-century Spanish galleon evolved from the carrack as a warship. Spain also needed large vessels to transport goods to and from her New World colonies. Mid-sixteenth-century galleons displaced between 300 and 600 tons; galleon capacity increased to 1200 tons by the end of that century. To appreciate the variety and ubiquity of the galleon design, consider the *Mayflower* and the *Vasa,* both galleons. The *Mayflower* was a small merchant galleon that famously landed pilgrims in the New World in 1620. The *Vasa* was a large Swedish warship that sank in Stockholm harbor on its maiden voyage in 1628; retrieved and restored in the 1960s she is today the only surviving galleon.*

The galleon was given a new hull form, longer and slimmer than the carrack. Whereas the ratio of hull to keel to beam length for the carrack was 3:2:1, the galleon length ratios were 4:3:1. Hydrodynamic drag was reduced, and the galleon was more maneuverable and seaworthy than the carrack. With a tumblehome bulge (the beam widest at the waterline, with the hull cross section tapering inward near the main deck; see fig. 3.3) the galleon was stable and difficult to board in combat. Stability was further improved in well-designed galleons by placing the large guns on two gun decks below the main deck and near the centerline to reduce heeling moment (of which much more in a later chapter). Galleons had a lower profile than earlier ships. The forecastle was reduced to permit a

*The *Vasa* sank for much the same reason as England's *Mary Rose* of the previous century: an overloaded prestige ship with gun ports too close to the waterline, she shipped water through these gun ports while maneuvering.

Figure 3.3. A model of the 1590 Dutch galleon *Gouden Leeuw*. This is an early galleon with a single gun deck (and stern chasers). Note the tumblehome shape and high stern castle. I am grateful to John Andela for providing these images.

spritsail for better handling and was moved a little further back so that the bowsprit could be placed in front (rather than sticking out of the top, as with carracks). This design change enabled galleons to sail closer to the wind. The stern castle, well-supported on a flat stern, was still quite high. This feature led to earlier galleons' being crescent-shaped and top-heavy, and so prone to capsizing in rough weather. The caravel's stern castle and overhanging stern deck became the integrated galleon quarterdeck and poop deck. The high stern decks required an internal tiller, extended by a whipstaff and tackle to the quarterdeck (the ship's wheel had yet to be invented), a characteristic of galleon design.

Galleons sported either three or four masts. The three-masters were rigged square on fore and main masts, and lateen on the mizzen. A fourth, bonaventure mast would be lateen-rigged. The square sails were wide but narrow-headed. An efficient arrangement of square topsails and topgallant sails adorned the main mast, and the bowsprit had its own spritsail. Sometimes a mizzen topsail was added. The general trend as galleons evolved was to increase the size and number of topsails and reduce the courses (lower sails). The English introduced recut sails that were flatter and more aerodynamic, sailing better to windward,* and the Dutch (around 1570) introduced telescoping topmasts that could be raised or lowered as circumstances dictated.

English galleons from the late sixteenth century were smaller and faster than their Spanish counterparts and with lower superstructures ("razed"—hence the term "race-built"). They were longer but more slender and more maneuverable, with improved rigging. Gunnery was emphasized more in the English navy, whereas the Spanish retained the old view that sea battles should be between boarding parties of soldiers. This difference in tactics, as much as the difference in ships, accounts for the results of the invasion of England by the great Spanish Armada in 1588. There was no single battle that decided this contest between galleon fleets, but rather a number of engagements and skirmishes over the course of a week.

*There must have been some backsliding in the Royal Navy regarding sails because, 250 years later during the Napoleonic war with France, Admiral Thomas Cochrane wrote: "I had ample opportunity of observing the superior manner in which the sails of the *Dessaix* were cut, and the consequent flat surface exposed to the wind; this contrasting strongly with the bag reefs, bellying sails, and breadbag canvas of English ships of war at that period" (quoted by Woodman).

Our interest here is more in the ships than in the course of the battle, so I will simply summarize what happened. King Phillip II of Spain wished to invade England for religious reasons and assembled a large armada (Spanish for "navy") off Calais, France. English fire ships scattered the Spanish vessels. When the armada attempted to reassemble, the Battle of Gravelines resulted. During this battle 11 Spanish ships were lost, with about 2,000 casualties; English losses were light. The Spaniards were driven northwards up the eastern coast of England, harried by the more maneuverable English ships (which had by this time run out of cannon shot, but the Spaniards weren't to know that). The armada struggled across the northern coast of Scotland, and then south past Ireland to home and safety. Only about half of the original 130 ships made it home; most of the losses were due to bad weather.

Only about 20 ships of the armada were true galleons, and over half were converted merchant ships. The English fleet consisted of about 60 smaller galleons; because they were faster and more maneuverable, they were able to stay upwind of the Spanish ships at Gravelines. English guns were used more effectively, hitting Spanish vessels, which were heeling over in the wind, below the waterline. The Spanish gunners were marines, trained to fight at close quarters. They could fire their cannons only once; they had not been trained to reload during a battle.* The English tactics were to stand off and fire, which their ships and cannons allowed them to do. The old method of naval warfare—boarding and hand-to-hand combat—would still be used into the eighteenth century by pirates in the Caribbean, and into the twenty-first century by Hollywood directors, but from the sixteenth century onwards real naval battles increasingly became engagements at distance, determined by superiority of firepower.

Tactics: Ships of the Line

The tactics employed by navies during the period covered in this chapter (the late 1500s to the mid-1800s) were dictated primarily by these circumstances: (1) ships were powered by the wind and (2) warships were

*Marine archeologists have found sunken Spanish galleons with most of their ammunition still on board.

armed to the teeth with increasingly devastating cannons.* Ships and cannons would evolve during these three centuries, but the wind remained constant—or rather it didn't—and the vicissitudes of wind and weather dominated naval formations and tactics.

We have seen that medieval sea battles mimicked those of antiquity in that they took place at close quarters. Ship A would ram ship B, whereupon the rammers would board the rammees (to coin a word) and then fight each other as if on land. Cannons changed all that, as we saw with the Spanish Armada in 1588. The introduction of cannons to warships led, over the century or so following the armada, to the adoption of line-of-battle formations by all the major European navies. Indeed, cannon armaments more or less obliged fleets of warships to form up in this way, or they would be seriously disadvantaged. Let me explain.

The structure of ships meant that most cannons were aimed out to the sides—port and starboard—of a warship, and so the ship was not much of a threat if approached from the bow or stern. The field of fire was limited; the cannons fired broadsides and could not angle their shots much forward or aft. The increasing range of cannons meant that, in a disorderly battle with friend and foe mixed up together, a friendly ship might get between you and your target, thus limiting your effectiveness in battle. To accommodate both these realities, ships formed a line of battle. One ship would take the lead, and the others would follow on behind, like ducklings following their mother. The battleline reduced a ship's vulnerability fore and aft because there were friendly ships in front and behind (unless the ship was at one end of the line), and the ship's broadsides could sweep only those areas of ocean that were empty of friendly ships. The simple geometrical logic of this formation was compelling, but its adoption was not so clear-cut. There were a couple of problems that needed to be solved before line-of-battle formations became really effective.†

First, for a number of ships to remain in line they had to be of roughly

*By the late seventeenth century gun decks on ships in the line of battle (*ships of the line)* became horizontal, rather than following the external lines and planking as on earlier galleons. This change permitted increased numbers of cannon.

†The first occasions in which both sides used line-of-battle formations occurred during the Anglo-Dutch wars of the late seventeenth century. The crew and firepower of opposing ships were roughly matched in quality; the lines quickly broke down chaotically into individual duels between ships.

equal sailing ability. Early in our period, naval fleets were composed of a few royal ships (i.e., purpose-built warships, often "prestige" vessels that were much better equipped than most) plus many other commandeered merchantmen. Maneuvers in line were tricky and slow: if tacking was called for, ships in the line would tack sequentially from the front, at the same place; otherwise the line would break up. Moreover, the line was a chain only as strong as its weakest link. Ships in the line had to be able to take punishment and stay in place, or the line would break. Both these factors led to a standardization of warships. Armed merchantmen became less useful; dedicated warships were needed for the line. By the seventeenth century, design standardization permitted the Royal Navy to introduce a rating system for ships so that commanders would know which ships were capable of sailing in the line and where they should be placed.* By 1690 an Anglo-Dutch fleet (the English and the Dutch were on the same side by this time) and its French enemy were able to engage each other with line-of-battle broadsides in the Battle of Beachy Head, though the lines quickly disintegrated (see fig. 3.4).†

Second, the ships in the line of battle had to be well organized, with an effective system of communication and with sufficient maneuverability to maintain the line in changing wind conditions. These lessons took decades to learn, but when learned they formed the basis for later eighteenth- and early-nineteenth-century naval warfare. In the optimal formation, lines formed across the wind, so that no ship took the wind from her neighbor's sails, and squared up with an enemy line, so that the lines sailed parallel and with the front and rear ships opposite each other to avoid doubling.‡ These configurations required careful and coordinated maneuvers. During an engagement, a ship would loosen the sheets so that some sails—those not furled—would spill wind and flap. These sails could quickly be brought into action if a sudden maneuver was ordered or was called for by the tactical situation ("let's get the hell out of here").

*The rating system itself evolved as ship size increased. Thus, in 1651 a first-rate ship of the line carried at least 60 guns. By 1680 the minimum requirement was 80 guns, and by 1700 it was 100 guns.

†The French won. The English commander, Torrington, fled, leaving his Dutch allies in the lurch. Later he needed to do some fast talking to avoid a court martial.

‡If line A was slightly in front of line B, the leaders of A could cross the front of B and double back, so that the leading ships of B would receive broadsides from both port and starboard—bad.

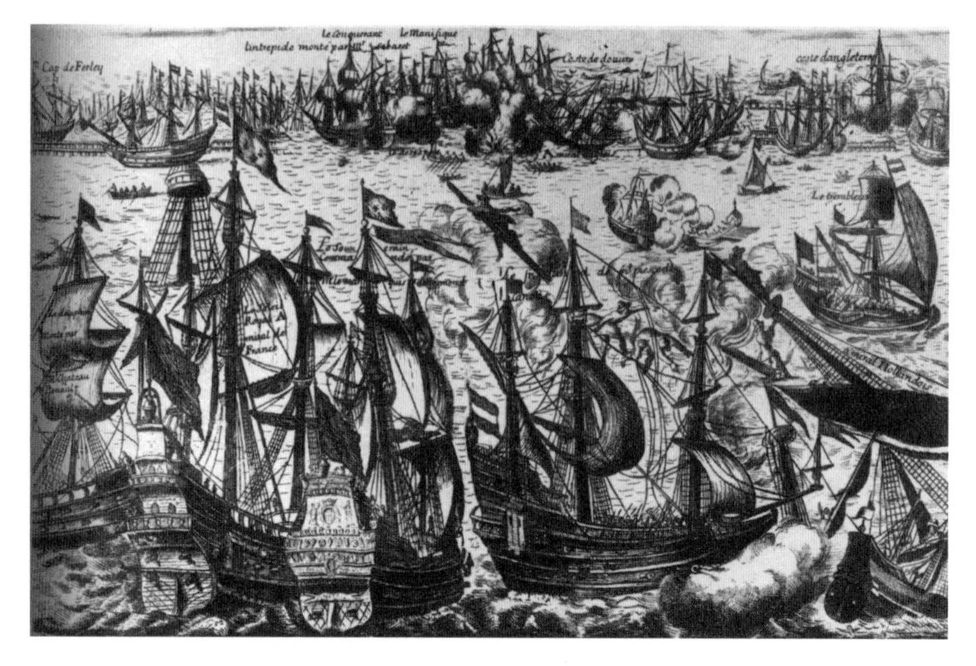

Figure 3.4. A confusion of galleons. This old print of the Battle of Beachy Head (1690) serves to show, if nothing else, how large sea battles could become confused in the early days of the Age of Sail. Image from Wikipedia.

Because opposing lines tended to form across the wind, one line was generally upwind of the other. Was this an advantage or not? The British and Dutch thought so; the French disagreed. The line that was upwind of its opponent had the tactical initiative; its commanders could choose whether or not to attack. Attacking upwind was much more difficult, given that ships of the line could sail only a couple of points to windward. As we saw at Gravelines, the leeward ships exposed their bottoms as they heeled and so risked getting shot "between wind and water." Another disadvantage for the contenders on the leeward side was that the huge amount of smoke generated by the gunpowder (Napoleon's "fog of war") blew in their eyes. Those in the windward fleet might not have been able to see the enemy clearly, but at least they could see each other. The English and Dutch, and any other naval commanders who desired a windward approach to battle, would often maneuver for days before coming to battle, in order to gain this advantage. They would fire broadsides when their ships were on a downward roll so that their shot would strike the enemy hull: a shot below the waterline would be a serious inconvenience.

The French favored the leeward side because they were often outnumbered and outgunned by their (usually English) enemy and liked to be able to withdraw.* Another advantage was that the lower gun ports were raised higher above the water as the ships heeled, whereas the windward ships heeled so that their lower gun ports were closer to the water. In rough weather this meant that the lower gun deck—with the heaviest guns—could not be used because of the risk of flooding. The French liked to fire their broadsides on the upward roll so as to destroy enemy rigging with chain-shot.† This enabled them to withdraw or outmaneuver the opposition because handling a ship under fire with damaged rigging was, to say the least, testing.

Effective or not, the French tactics caused fewer casualties than the English or British‡ because most of the crew would be located on the main deck or gun decks, rather than up in the rigging.

The French and the British were at each other's throats constantly during the strife-torn eighteenth century. Sea battles of this period were often indecisive because the French were wary of the size and firepower of the British fleets and would engage only cautiously and withdraw

*The inability of the windward opponent to pull out of a battle if things were going badly was exploited by Lord Howe in the Battle of the First of June. This battle, which took place in the Atlantic in 1794, was the first sea battle of the French revolutionary wars. Confident of the superiority of his own ships over those of his French opponent, the British commander deliberately broke line and, being to windward, was able to quickly pass some of his ships through the enemy line to re-form on the leeward side, resulting in a decisive victory. Twelve-year-old midshipman William Parker reported: "The smoke was so thick that we could not at all times see the ships engaging ahead and astern. Our main-topmast and main yard being carried away by the enemy's shot, the Frenchmen gave three cheers, on which our ship's company, to show they did not mind it, returned them the three cheers, and after that gave them a furious broadside" (Lewis).

†Cannons fired several kinds of shot: Round-shot consisted of single cannonballs, used to smash hulls. Chain-shot was two small balls connected by a chain, fired into the rigging at short range. Grape-shot consisted of a large number of small musket balls loosely held together. When fired, the grape-shot would disintegrate, with the musket balls emerging from the barrel individually, so that the cannon became a giant shotgun. Grape-shot was intended to kill people and, in particular, to sweep the enemy decks.

‡I have used the term "English" for the period before the 1707 union with Scotland. Because we are now well into the eighteenth century, I hereafter use the term "British."

when it seemed like a good idea to do so. To counter this indecisiveness, which benefited nobody, a number of innovations were brought in during the American Revolution. One of these was the carronade, a large-caliber, short gun of limited range but of devastating force. The main advantage of this new weapon, first constructed in Scotland, was its light weight. The carronade was less than half the weight of the long guns; as a result it could be placed on the forecastle and quarter deck and directed forward or aft. (Placing heavy cannon on these high decks would make a ship top-heavy and unstable.) A second innovation led to increased field of fire for the main guns; they could traverse a wider angle and so deliver broadsides over a greater span of ocean. A third improvement of this period—copper sheathing on the underside of the hull—was directed against wooden ships' universal enemies, shipworms and crustaceans, and enabled ships to stay at sea longer.

Tactical innovations during this period, and somewhat later in the Napoleonic wars, also helped to break the deadlock. The Battle of the First of June, already mentioned, was the result of a commander's breaking his own line—normally the safest formation. An earlier British commander, that dashing, intemperate gambler Admiral George Rodney, in an earlier war (the American Revolution) similarly took advantage of a change in wind direction to break his own line and sail through that of his French opponents, at the Battle of the Saintes off Guadeloupe in 1782. His raking fire led to a significant victory.* This was the era of British Royal Navy dominance (arguably from the mid-eighteenth century to World War I). Its greatest commander, Lord Nelson, himself introduced new tactics, most famously at the Battle of Trafalgar. Nelson split his line in two, forming a windward column with his flagship HMS *Victory* in front and a leeward column led by his able deputy, Collingwood, in the *Royal Sovereign* (both these ships were 100-gun giants). The two columns penetrated and passed through a ragged line of French and Spanish ships, taking hits on the way in and returning them with interest as they passed through. The imbalance in firepower resulting from this tactic is a simple geometrical consequence of lines' firing broadsides. The tactic became known as crossing the T and has been picked over by military analysts ever since the 1805 battle. Given that the result of the

***Raking* fire is directed along the length of a ship and so has greater chance of finding a target on the decks. It is the naval equivalent of *enfilading* fire on land.

battle was a smashing victory for the British that eliminated the threat of Napoleon's invasion of England and eclipsed both the French and the Spanish navies, you will be unsurprised to learn that the analysts conclude that Nelson got it right.[1]

The innovations I have described here so far refer to large-scale tactics, of nations or fleets. At the smaller scale of individual ships there were always, throughout the Age of Sail, duels that required innovation and improvisation, as hard-pressed captains battled each other with wits and seamanship as well as with broadsides. Here is Admiral Cochrane recalling his days as commander of the small brig HMS *Speedy* in 1801, which was sighted by an enemy frigate:

> On the 18th we again put out to sea, and towards evening observed a large frigate in chase of us . . . To cope with a vessel of her size and armament would have been folly, so we made all sail away from her, but she gave instant chase, and evidently gained on us. To add to our embarrassment, the *Speedy* sprang her maintop-gallant-yard, and lost ground whilst fishing it. At daylight the following morning the strange frigate was still in chase, though by crowding all sail during the night we had gained a little on her; but during the day she again recovered her advantage, the more so, as the breeze freshening, we were compelled to take in our royals, whilst she was still carrying on with everything set. After dark, we lowered a tub overboard with a light in it, and altering our course thus fortunately evaded her. (Woodman; also quoted in Lewis)

Here is Captain Sir Edward Berry, talking to Nelson during the battle of Trafalgar: "I will do the utmost, my lord; get the engines to work on the sails—hang butts of water to the stays—pipe the hammocks down, and each man place shot in them—slack the stays, knock up the wedges and give the masts play—start off the water Mr James, and pump the ship." Berry's meaning is explained as follows: " 'Engines' are pumps with which to wet the sails, since damp sails set fairer and will not catch fire in a fight; water butts on the deck are further fire precautions; shot placed in the windward hammock netting on a deck helps balance the ship and a level ship sails faster; the slackened stays and masts given play both allow more sail to be set; pumping the ship and draining the water butts lightens the load." (Both quotes are from Nicolson.)

Another tactic, brought to bear during a close-quarter engagement, when the opposing guns were so close to each other that they could touch, was to direct the cannon fire upward and downward. Guns on high decks would fire down and those on low decks would fire up, so that several enemy decks would be devastated. At such close ranges, powder was reduced to slow down the shot so that it would remain within the enemy ship and not blast out of the other side, perhaps damaging a friend.

Even at the peak of its strength and influence, however, the Royal Navy did not have things go all its own way. During the curious War of 1812* the young U.S. Navy took on the vastly larger Royal Navy and acquitted itself well. The United States possessed far fewer ships and none larger than a frigate—but what frigates!† Given the disparity of numbers, the young republic's tactics were, sensibly, to avoid full-blooded battle (and naval battles in the Age of Sail were very bloody) and instead to harry British merchant shipping and pick off Royal Navy ships only under favorable circumstances. To this end, fast and powerful frigates were constructed specifically to outperform their British counterparts. Smaller and faster than the Royal Navy's ships of the line, their hit-and-run approach kept them out of trouble and ensured that they were at an advantage when engaging their chosen opponents.

Early results were quite a blow to the Royal Navy's prestige as well as to its ships. Thus, in 1812 the USS *Constitution* took on HMS *Guerriere* off the coast of Nova Scotia and whacked her in 30 minutes. Both ships were frigates, but the *Constitution* had 44 guns to the *Guerriere*'s 38. In the same year, south of the Azores the USS *United States* (44 guns) pul-

*A silly little war between Britain and the United States that was fought rather half-heartedly on land, though not at sea. Depending on your point of view, the war was fought either to free up international trade or to protect Canada. Either way, the reasons for it evaporated with the defeat of Napoleon in 1815, and the belligerents stopped fighting.

†Frigates in the Age of Sail were three-masted and fully rigged, typically with 30–40 guns. They were too small to act as ships of the line; instead, they were very effective scouts and escort vessels. The U.S. "super" frigates were bigger than their British counterparts and were particularly well-built, usually of live oak, a native American wood—one of the densest and hardest in the world, becoming even harder when milled and left out to weather.

verized HMS *Macedonian* (38) at long range.* U.S. crews were much better trained in gunnery than most of the Royal Navy's enemies, and their seamanship was second to none. A series of such early defeats for the British (also in 1812 the *Constitution* beat HMS *Java* in three hours, off the coast of Brazil) was not only a slap in the face; it also prompted a rethinking about strategy. Instead of standing orders to engage enemy vessels on sight, Royal Navy captains were instructed *not* to engage U.S. frigates unless the British ships had a distinct numerical advantage. Some later results restored Royal Navy pride,† but the message to the navies of the world was clear—there is a new kid on the block.‡

Two Old Ladies

My goal is to tell you about the evolution and science of sailing ships in general, but I can't resist including this anecdotal aside about two particular sailing vessels because they carved for themselves such impressive histories and because both are still with us. These grand old dames—one still sprightly and the other now infirm—have many admirers who visit them each year and are both still very photogenic. If their stories don't stir your blood, then you are reading the wrong book.

HMS "Victory"

The world's oldest commissioned warship, the HMS *Victory,* was launched in 1765 though not commissioned until 1768 (fig. 3.5). It is reckoned that weathering the *Victory*'s timbers for three years before her commissioning accounts for her long life of active service. An estimated 6,000 trees went into her construction. Her hull is 2 ft thick at the waterline. She

* "It was soon ascertained that the enemy had shot ahead to repair damages, for she was not so disabled but she could sail without difficulty; while we were so cut up that we lay utterly helpless. Our head braces were shot away; the fore and main topmast were gone; the mizzen mast hung over the stern, having carried several men over in its fall: we were in the state of a complete wreck."—Samuel Leech, *Macedonian* fifth gun crew (quoted in Lewis).

† For example HMS *Shannon* (38 guns) defeated USS *Chesapeake* (38) off Boston harbour in 1813. The *Chesapeake*'s fir timbers were later used to make a mill in England, which still stands.

‡ "We have an infant navy to foster, and to organize, and it must be done."—Captain Thomas Truxton, USN, 1798 (quoted in Lewis).

Figure 3.5. First-rate ship of the line HMS *Victory*. Thanks to Mike Cawood for these images.

Table 3.1 Vital statistics for HMS *Victory* and USS *Constitution*

	HMS *Victory*	USS *Constitution*
Type of vessel	Ship-of-the-line warship	Super frigate
Dates in service	1768–1812	1797–1881
Construction	Mostly English oak	Live oak
Size		
Displacement	3,500 tons	2,200 tons
Length	227½ ft	202 ft, including bowsprit
Beam	52 ft	43½ ft
Draft	24½ ft	14 ft
Main mast	205 ft above waterline	220 ft with top, topgallant, royal mast
Bowsprit	110 ft long	65 ft long
Main yard	102 ft across	95 ft across
Rigging	26 miles	15 miles
Total sail area	60,000 ft^2	43,000 ft^2
Number of crew	820	450
Armament	104 guns on three decks	44 guns on two decks
Maximum cannon range	2,600 yards	1,200 yards

was decommissioned in 1812 and now can be seen in dry dock in Portsmouth harbor on the southern coast of England, where young navy personnel will tell you everything you could possibly want to know about her. Table 3.1 summarizes a few of her vital statistics.*

Victory could achieve a maximum speed of about 10 knots and had a reputation in her day for excellent sailing qualities. Her 104 guns included thirty 12-pounder carronades on the upper gun deck, twenty-eight 24-pounder cannons on the middle gun deck, and thirty 32-pounder cannon on the lower gun deck. The heaviest guns were on the lowest deck, and they had a maximum range of about 2,600 yards (400 yards was considered point-blank range). The quarterdeck held twelve

*The complexity and size of first-rate ships of the line made them "undoubtedly the most evolved, with the most elaborate ordering of parts, the world had ever seen" (Nicolson).

12-pounder carronades, and the forecastle two 12-pounder cannons and two 68-pounder carronades.

HMS *Victory* fought against the French during the American Revolution, in the second Battle of Ushant in 1781. She fought again in the Battle of Cape St. Vincent (1797), off the Portuguese coast, against a Spanish fleet which was, at that time, allied to revolutionary France. In both cases the British ships of the line were outnumbered by their enemies (the Spanish ship of the line *Santissima Trinidad* was even bigger than the *Victory*, with 136 guns), and in both cases *Victory* contributed to significant victories. She took a breather during 1800–1803, when she was refitted. Her finest hour was undoubtedly at Trafalgar, where she crippled the French flagship *Bucentaure* and grappled with *Redoutable*. The *Redoutable* eventually struck her colors to the *Victory*, but not before a marksman had mortally wounded Admiral Nelson (a plaque onboard the *Victory* now commemorates the spot). In all, 57 crew of the *Victory* were killed and 102 wounded.* Today only one of the original sails that she deployed at the battle of Trafalgar still exists. This fore topsail was placed on display during the Trafalgar bicentennial in 2005: it weighs 800 pounds and has 90 shot holes in it.

USS "Constitution"

The world's oldest floating commissioned warship, the USS *Constitution* is one of six super frigates ordered by Congress in 1794 (figs. 3.6 and 3.7). In service from 1797 until 1881 she has, since retiring from active service, resided in Boston harbor. The *Constitution* was strongly built of live oak, with planks up to 7 in. thick. Her vital statistics (see Table 3.1) make for an interesting comparison with those of the *Victory*, and emphasize the difference between a large ship of the line and a frigate. Her 24-pounder cannons each weighed 3 tons and, it is claimed, could blast

*Midshipman R. F. Roberts was on board during the action at Trafalgar. "The rascals have shot away our mizen mast and we are very much afraid of our main and foremasts. The *Royal Sovereign* has not a stick standing—a total wreck . . . You can have no conception whatsoever what an action between two fleets is; it was a grand but an awful sight indeed; thank God we are all so well over it." Again: "It was as hard an action as was ever fought. There were but three alive on the quarterdeck, the enemy fired so much grape and small shot from the rigging, there was one ship so close we could not run out our guns the proper length" (Lewis).

Figure 3.6. The frigate USS *Constitution*. At over 200 years old, she has been refitted, renovated, and restored so many times that, apart from the keel, very little of the original vessel survives. (The helm was destroyed in Old Ironsides' fight with HMS *Java*, so *Java*'s wheel was afterwards appropriated for use in the *Constitution*.) U.S. Navy photo.

through 2 ft of wood at 100 yards; their maximum range was 1,200 yards. She was also armed with 32-pounder carronades.

Her first tour of duty took *Constitution* to the coast of North Africa, where she served in action against the Barbary corsairs, who were demanding tribute for granting the United States access to Mediterranean ports. She blockaded ports and bombarded forts. *Constitution* made her reputation in the War of 1812. In that year she evaded British squadrons while seeking out enemy targets: the first such target, located in the Gulf of St. Lawrence, was the *Guerriere*.* At one point during the brief engagement, round-shot from the Royal Navy frigate bounced off the re-

*The *Constitution*'s captain Isaac Hull reported: "Immediately made sail to bring the ship up with her, and 5 minutes before 6 p.m. being along side within half pistol shot, we commenced a heavy fire with all our guns, double shotted with round and

Figure 3.7. Left: Another view of USS *Constitution. Right:* The *Constitution* in dry dock, showing off her copper bottom. U.S. Navy photos.

silient sides of *Constitution*, earning her the nickname "Old Ironsides," which has stuck to this day. The U.S. victory in this engagement was a great boost to morale.

In the same year the *Constitution* was involved in a more drawn-out battle with HMS *Java* off the coast of Brazil: this was her toughest fight

---

grape, and so well directed were they, and so warmly kept up, that in 15 minutes his mizzen-mast went by the board, and his main yard in the slings, and the hull, rigging and sails very much torn to pieces. The fire was kept up with equal warmth for 15 minutes longer, when his main-mast and fore-mast went, taking with them every spar, excepting the bowsprit . . . After informing you that so fine a ship as the *Guerriere*, commanded by an able and experienced officer, had been totally dismasted, and otherwise cut to pieces, so as to make her not worth towing into port, in the short space of 30 minutes, you can have no doubt of the gallantry and good conduct of the officers and ships company I have the honour to command" (Lewis).

against a faster opponent. Teddy Roosevelt wrote a spirited account in his *Naval War of 1812:*

> The stump of the *Java*'s bowsprit got caught in the *Constitution*'s mizzen-rigging, and before it got clear the British suffered still more. Finally the ships separated, the *Java*'s bowsprit passing over the taff-rail of the *Constitution;* the latter at once kept away to avoid being raked. The ships again got nearly abreast, but the *Constitution,* in her turn, forereached; whereupon [U.S.] Commodore Bainbridge wore, passed his antagonist, luffed up under his quarter, raked him with the starboard guns, then wore, and recommenced the action with his port broadside . . . The great superiority of the Americans was in their gunnery. The fire of the *Java* was both less rapid and less well directed than that of her antagonist; the difference of force against her was not heavy, being about as ten is to nine, and was by no means enough to account for the almost fivefold greater loss she suffered.

Later the *Constitution* captured eight smaller Royal Navy vessels, including two in one engagement off Madeira in 1815. "Old Ironsides" ended the war undefeated. In the years 1821–1828 she served in the Mediterranean. After a refit she continued in active service for several decades more, at one point prowling the Atlantic coast of Africa intercepting slavers bound for the New World. Today her mission is to promote the U.S. Navy, whose personnel still man her.

Dutch Treats

From galleons to ships of the line I have taken you right through the mainstream of the evolution of Age of Sail warships. There were differences in design, and within a given category of ship, there were national variants. Thus, an English ship tended to carry more guns than her French counterpart (and enemy, usually, in the Age of Sail). So, the English vessel would sail deeper and may have been unable to open the lower gun ports in heavy weather. (For this reason the French would remove some guns from captured English vessels before sailing them.) On the other hand, French vessels suffered from a surfeit of useless baroque decoration, which hampered seaworthiness. In the early years of the Age of Sail, the French designs were better, on the whole, and provided the pattern for the nineteenth-century fighting ships that we have already encountered.

What about commercial vessels in the Age of Sail? Here we must turn to the Netherlands because this small nation, born at the beginning of our period, was hugely influential in the design of merchant vessels. It was the merchantmen of the seventeenth through the nineteenth centuries, rather than the warships, that contributed most to the look of our modern sailing craft, and so we should delve a little deeper into the Dutch contribution. After a long occupation by Hapsburg Spain, the Dutch gained their independence in the seventeenth century and immediately set about establishing a trading empire. Baltic trade led the Dutch to develop the *boyer* (*boeijers*), a small vessel of under 100 tons that was required to sail to windward a lot of the time. So the boyers adopted a modified lateen sail, with the yard cut off in front of the mast to save space—and the gaff rig was born. Gaff sails hang behind the mast and

Figure 3.8. The Dutch introduced the gaff sail as an improved lateen to help trading vessels sail to windward. This ship has a mixture of a square-rigged foremast and mainmast, with a fore-and-aft gaff sail and triangular staysails. This combination is quite common in modern tall ships (see also the *Constitution*'s rigging in fig. 3.6) and we might regard it as a hybrid or transition rigging, partway between the old-fashioned square-riggers and modern fore-and-aft sailing vessels. Thanks to Darillo for this image.

Figure 3.9. A schooner. This old-style gaff-rigged sailing ship, with all sails of the fore-and-aft type, became very popular for merchant vessels in the nineteenth century. Thanks to Simona Manca for this image.

swing symmetrically to port or starboard (see figs. 3.8 and 3.9). Thus, they tack equally well in both directions. Because the gaff sails are easier to work than the lateen, the rigging is simpler and the vessel requires fewer crew. Crew size is important for commercial viability in a competitive environment. The gaff sail had other advantages over the lateen. It could be furled easily and quickly by drawing it to the mast, like a curtain. It freed up space in front of the mast, which then permitted deployment of staysails, triangular sails rigged between masts. Of great importance to later pleasure boat design, the boyers also carried jib sails, staysails running from fore mast to bowsprit.

Another Dutch contribution to commercial vessels was the buss, a shallow-draft fishing boat with hinged masts that could be lowered to reduce leeway when the crew was working the nets. To accommodate the lowered masts, the hull of the buss had to be long, and long hulls became a characteristic of Dutch ships in general.* The great length-to-

*An incentive for trading vessels to be narrow came from port duties, which often were higher for ships with larger beams.

beam ratio of another Dutch design (5:1, later 6:1) gave rise to its name: flute (Dutch *fluit*), from a resemblance to the slender wine glass. The flute had a low forecastle and a high galleon-like stern, comprising half, quarter, and poop decks, tapering to a point at the stern. Their length meant that flutes were slow to turn, but they were generally reckoned to be good sailers and stable ships. Their main commercial asset was that they required a very small crew: eight for a vessel carrying 150 tons of cargo. This small complement was due in large part to very efficient rigging. Flutes deployed large topsails and led the general trend to larger topsails relative to courses as time went by. The advantage of a large topsail was that in light winds the other sails could be furled, improving visibility. Plus, buildings around harbors often robbed lower sails of wind, and so topsails were useful when maneuvering in harbors.

Flutes were a very successful class of merchant ship. Introduced around 1600, flutes were made in five or six variants by 1650; two decades later, 10,000 flutes plied their trade across the oceans of the world. Another pointer to the future: flutes were the first trading vessels with fully specified designs. Standardization (despite the variants) reduced production costs significantly: the cost of manufacturing a flute was 40% of the cost of producing the earlier carrack. Flutes were mass-produced in the Zaanstreek shipyard near Amsterdam. To cut costs they were made of pine, rather than the more durable oak, and so lasted only about 20 years.

The Dutch also contributed to warship development. As we saw in figure 3.1, they built galleons in the seventeenth century. A later development, the pinnace, was a light frigate resembling a mini-galleon. An even lighter warship was the *jachtschip* ("hunting ship"). In the late seventeenth century the Dutch presented an elegant jachtschip as a gift to King Charles II of England, thus giving us the name—yachts—for our present-day pleasure craft.

East Indiamen and Others

In the seventeenth and eighteenth centuries several western European maritime trading nations established colonies or trading monopolies with peoples in South and East Asia. Luxury trade goods such as spices, silks, works of art, and, increasingly, tea were brought from India, China,

Japan, and Indonesia to be sold in the markets of Europe. The Dutch, French, and British granted trading monopolies to their own East Indies companies, who operated very profitably though inefficiently as a consequence. The British came to dominate this trade by the nineteenth century, by which point the specially commissioned merchant vessels—the East Indiamen—that transported their cargo across the world had become very large and prestigious vessels. With ornate interiors and gilded carving, they were the result as much of cozy monopoly as of maritime evolution. The ending of the monopolies in the early nineteenth century spurred evolution, as did American independence, resulting in even larger and more impressive merchant ships, as we will see.

But I am getting ahead of my story. From the seventeenth to the nineteenth centuries, merchant ships became better and better as new techniques and equipment were put into practice. By the end of the seventeenth century the high sterns of merchantmen had been reduced, and the ships' waists bridged by gangways at the side, thus presenting a more level profile. Aerodynamic drag was consequently reduced, and so was the tendency to heel over in a crosswind. Frame construction improved, with overlapping joints on the strakes. The improvements of the seventeenth century became more widespread in the eighteenth as their benefits became evident, and new improvements were added. Copper sheathing replaced pitch and tar, as we have seen. Also in the eighteenth century maneuverability was improved with the introduction of the wheel, which replaced the old steering lever used in galleons.

Most of the improvements occurred in the rigging. Fully rigged ships grew more and more sails. The bowsprit was extended by adding a jib boom to permit extra jib sails. Staysails proliferated in the eighteenth century to cover all the stays on a ship. Topsails grew larger than mainsails. Skysails were added to the fore and main, and royal sails to the mizzen masts. Stud sails appeared, like wings. In the 1600s the yards of a fully rigged ship would get smaller with height above the deck, but this difference lessened as the topsails (lower topsail, upper topsail, topgallant, royal and sky sails) became more important. We have seen that much of the incentive for improved rigging efficiency came from merchants' desire to reduce the crew size of their vessels, and hence the transportation costs of their trade goods. We have also seen that—even for a good, efficient, and well-designed ship—there were miles and miles

Figure 3.10. Rigging—miles of it, both standing and running: stays, shrouds, lifts, sheets, tacks, halyards, jeers, ratlines, blocks. It must have been essential yet almost impossible to keep it all ship-shape. Thanks to Bruno Girin / DHD Multimedia Gallery and to Mike Cawood for these images.

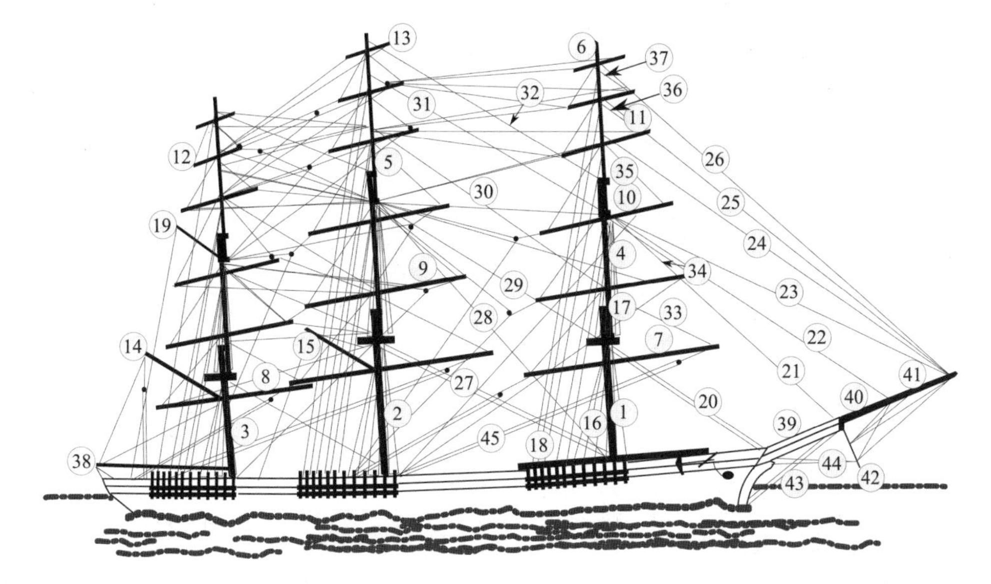

Figure 3.11. Standing and running rigging of a fully rigged ship (not every yard and stay is labeled): (1) foremast, (2) mainmast, (3) mizzenmast, (4) topmast, (5) topgallant mast, (6) royal and skysail masts, (7) fore yard, (8) cross-jack yard, (9) lower-topsail yard, (10) upper-topsail yard, (11) topgallant yard, (12) royal yard, (13) skysail yard, (14) spanker gaff, (15) trysail gaff, (16) lower shrouds, (17) topmast shrouds, (18) backstay, (19) monkey gaff, (20) forestay, (21) fore-topmast stay, (22) jib stay, (23) outer-jib stay, (24) fore-topgallant stay, (25) fore-royal stay, (26) fore-skysail stay, (27) mainstay, (28) main-topmast lower stay, (29) main-topmast upper stay, (30) main-topgallant stay, (31) main-royal stay, (32) main-skysail stay, (33) lift, (34) lower-topsail lift, (35) upper-topsail lift, (36) topgallant lift, (37) royal lift, (38) spanker boom, (39) bowsprit, (40) jib boom, (41) flying jib boom, (42) dolphin striker, with martingales, or stays running forward to jib boom and flying jib boom, (43) bobstays, (44) back ropes, (45) braces.

of rigging (see Table 3.1 and figure 3.10). Figures 3.11 and 3.12 show the names and locations of the running and standing rigging and of the sails of a fully rigged ship.

So the East Indiamen that plowed through the oceans from India and China around the Cape of Good Hope to London, or from Indonesia to Amsterdam, were large, fully rigged ships (typically with three masts, two of them square-rigged and one a jib sail) and were faster than their predecessors. They were also large, at 500 tons, and growing in both size (to an average of 1,200 tons by the year 1800) and numbers. The East

Figure 3.12. Fully rigged ship under sail: (1) foresail, (2) mainsail, (3) crossjack, (4) spanker, (5) lower topsail, (6) upper topsail, (7) topgallant sail, (8) royal, (9) skysail, (10) fore-topmast staysail, (11) jib, (12) outer jib, (13) flying jib, (14) main-topmast lower staysail, (15) main-topmast upper staysail, (16) main-topgallant staysail, (17) main-royal staysail, (18) mizzen staysail, (19) mizzen-topmast staysail, (20) mizzen topgallant staysail, (21) mizzen-royal staysail.

India companies were founded around 1600, and the era of the East Indiamen encompassed the two centuries from 1620 to 1830. These ships were almost as heavily armed as warships because they carried valuable cargo through pirate-infested waters. Some carried as many as 60 cannons, though of small size compared with those of warships. Because of their armaments, East Indiamen had the tumblehome shape of warships.

I have mentioned the prestige attached to these merchant ships. As well as ornate decoration they carried wealthy and influential passengers in comfort, accommodated on two decks within the hull. Because of the passenger accommodation and the armaments, the hulls were not tapered near the stern. Although a wide stern was necessary to support the extra weight—East Indiamen were among the largest ships of their day—it was hydrodynamically inefficient and slowed down the ships. The length-to-beam ratio of 4:1 also slowed them down compared with later vessels, and so, even though East Indiamen were speedier than other merchant vessels of their period, they could make at most one return trip per year from Europe to Asia, limited by the prevailing trade winds.

Two factors combined to end the dominance of East Indiamen as the super freighters of the nineteenth century: U.S. independence and the end of the East India companies' monopolies. The British parliament terminated the cozy monopoly of the "Honorable East India Company" on trade to India and China in 1833. This move immediately exposed the merchant ships to market forces. East Indiamen were phenomenally expensive to built and to run; the cost per ton of cargo was much higher than it needed to be, propped up by inefficient practices that had been insulated from competition during the period of monopoly. Merchants in the United States were well aware of these inefficiencies—and of the opportunities they presented. The young republic was eager to expand markets and needed to build large cargo ships that could trade around the world, competing with the large British merchant fleet but without the protection of the British navy. The vessels that these U.S. merchants came up with to lever themselves into the lucrative Southeast Asian market were the clipper ships, probably the best and the most beautiful of sailing ships.

Clippers—Yankee and Tea

In addition to the opening up of the South and East Asian market to competition, there were other incentives to American overseas commerce. The discovery of gold in California in 1848 stimulated trade from New York to San Francisco, around the Horn. This long journey was very profitable, and fast clippers could pay for their construction costs in just one trip. Gold discoveries in Australia in 1851 led to a similar "demand for supplies." Concurrent with free trade incentives with the British Empire —the Navigation Laws were repealed, permitting U.S. ships to deliver, for example, tea from China to London—these economic developments over a 20-year period provided powerful motivation for American shipbuilders to design and construct sailing vessels built for speed, not comfort, that could travel the oceans of the world with valuable cargo. The clippers would be much more economical to run than the old East Indiamen because of their speed and because they were not designed to carry

Figure 3.13. Model of a Baltimore clipper. A fast blockade runner, smuggler, and privateer ship from the early 1800s, this ship was the forerunner of the great Yankee clippers of the 1850s. Thanks to John Andela for this image.

passengers or heavy guns—they would be far too fast to be inconvenienced by pirates. Another benefit of speed: perishable cargo such as tea would command a higher price if delivered fresh to the traders in London, as well as for being the first delivery of the current harvest of tea leaves.

The United States had something of a head start in the speed stakes. In the years around 1800, shipbuilders along the eastern seaboard had produced a small vessel that was designed for speed. The Baltimore clipper (fig. 3.13) was produced in numbers and was a favorite of American privateers during the War of 1812. Called "clippers" because they clipped off the miles at a great rate, these ships were square-topsail schooners with tall raked masts carrying acres of sail and with the narrow hulls that are the hallmark of ships built for speed. Baltimore clippers ran British blockades and later ran guns and other war munitions to South America, where Spanish colonies were struggling for independence. In more peaceful pursuits, Baltimore clippers were commonly seen trading coffee, sugar, cotton, and flour between the United States and many of the European colonies in the Caribbean.*

The clippers grew larger and then much larger, with hull shapes increasingly slender. In 1815 a clipper might have a hull length-to-beam ratio of 4:1 and displace 500 tons; by 1840 her successor would be 5½:1 and displace 1,200 tons. The first American clipper to muscle in on the Asian run to London shocked British merchants by racing from Hong Kong in only 97 days—three times quicker than the East Indiamen—and gained a foothold for U.S. vessels in the formerly closed East Asian trade.

And not just in the East Asian trade. Journey times across the Atlantic fell dramatically. In 1620 the *Mayflower* took 66 days to cross; by 1820 this passage required only 23 days, and fifteen years later was further reduced to 14 days. In their heyday in the 1850s clippers became the fastest commercial sailing vessels ever built.† With average speeds of up to 20 knots for some voyages, they could sustain 35 knots over a 12-hour period. Over long distances the speeds were lower, but still much quicker than anything that went before. The 1,700-ton *Flying Cloud,* a famous American clipper, broke the speed record for the 18,000-mile trip from New York to San Francisco by completing this journey in 89 days. A latecomer, the 920-ton British clipper *Cutty Sark* (famous for being the only surviving clipper ship, albeit in dry dock), sailed from London to Sydney, Australia, in 73 days and from Shanghai to London in 110 days.

The long, sleek clipper ships had sharp bows and rounded (rather than square) overhanging sterns. (In chapter 6 we will see why these characteristics led to high hull speed.) Towards the end of the clipper era—which lasted only a generation—the hulls were built with iron frames, covered with wooden planks. Typically clippers had three masts, each adorned with up to five yards—plus staysails, plus stunsails (stuns'ls), etc. Clippers ranged from about 500 to 4,555 tons, with the largest being the *Great Republic,* launched before a cheering crowd of 30,000 in East Boston in 1853. Boston and the eastern seaboard boatyards produced these revered American ships (hence their name, "Yankee clippers").

*Less creditable cargo of fast clipper ships included opium and slaves. Smaller clippers were popular with smugglers and West Indian pirates.

†*Lightning* set the all-time record for distance sailed in 24 hours: 436 nautical miles. The radically streamlined hulls of such extreme ships sacrificed cargo size for speed: the prows were sharply pointed.

The British recovered from their initial shock over the speed and competition from the American clippers and, keen to regain market share, got in on the clipper business. Eventually they built roughly the same number as the Americans. British-built "tea clippers" were smaller on average (between about 450 and 950 tons) and were narrower in the beam.* They were faster than their American competitors in light weather but were not so good when the going got rough. Direct competition for the Chinese and Indian tea trade often took the form of races, with bets being placed in London on which ship would be the first to make it to the docks. U.S. merchants lost interest in the tea trade after 1855, but competition between them and the British—and between compatriot clippers—continued for another decade or so.

The most famous clipper race occurred in 1866. This event was not an officially sanctioned, formal race, with a starting line and a starting pistol; but, as often happened with the tea-clipper voyages, it quickly turned into a race across the world. It happened like this. The Chinese port city of Fouchow opened for trade with the West. Fouchow was closer to the tea-producing regions of China, and so the tea was loaded onboard the eagerly waiting fleet of clippers while fresh.† The first ship that docked in London could anticipate premium prices for her cargo. The "tea season," when clippers left Chinese ports bound for London, was late May to early June. Sometimes the ships were in such a hurry that paperwork was left incomplete. The clippers would race through the South China Sea, across the Indian Ocean, around the Cape, up the South Atlantic, past the Azores and up the North Atlantic, and then eastward along the English Channel, and northward again to enter the river Thames. Crews (typically 40 strong) were proud of the ships they worked and of their accomplishments in racing them. They were paid top rates, with bonuses for getting home first.

On 28 May 1866 no fewer than ten clippers left Fouchow, loaded with tea. The race was on, and eager crowds followed the news at it progressed. *Taeping, Fiery Cross,* and *Serica* were the fastest out of the

*The British clipper *Thermopylae,* great competitor of the *Cutty Sark,* was 212 feet long and 36 feet across the beam, so that her length-to-beam ratio was 5.9:1. The U.S. giant *Great Republic* was exceptionally narrow: at 325 feet long with a 53 feet beam, her ratio exceeded 6.1:1.

†Cargo was tightly packed, not only to increase the amount transported but also to prevent shifting during transit.

blocks, with *Ariel* closing. Telegrams were sent around the world to interested punters as the ships passed by ports en route. Crowds massed at the docks to wave them on; the masts and bulwarks of each ship were painted differently, so each was readily identifiable. Across the Indian Ocean the leading four were often within sight of one another. On 29 August they were dead level at the Azores, piling on sail and heading for home. *Ariel* and *Taeping* pulled away under full sail in the English Channel. Betting continued in London—and onboard as well: the crews of *Fiery Cross* and *Serica* had bet each other a month's pay on the outcome of their race. In the Thames estuary the two front-runners were neck and neck. Agonizingly for the *Ariel* crew, *Taeping* was picked up by a faster tug and so arrived in dock 20 minutes ahead—after a 99-day voyage. *Serica* came home later, on the same tide, and *Fiery Cross* within another two days. After some deliberation, the "race" for first place was declared to be a dead heat.

Sadly, the days of clipper races were numbered. Within a decade, most of these magnificent ships would be scrap. Such is the march of progress and the economics of long-distance trade. The *Great Republic* sprang a leak off Bermuda in 1872 and was abandoned. *Flying Cloud* was condemned and sold in 1874. Fastest of them all, the Yankee clipper *Lightning* shone brightly but all too briefly: launched in 1854 she burned while loading wool in Australia, in 1869. *Ariel* probably foundered in the Southern Ocean in 1872: only an empty lifeboat was found. *Taeping*'s life was even shorter: launched in Greenock, Scotland, in 1863, she was wrecked on a reef in the China Sea in 1871, on her way to New York.

The real cause for the clippers' demise was not rough seas, but rough economics and technological progress. In some ways they were victims of their own success. So good were they that too many were built, and even as early as 1855, the freight rates they charged were dropping. The Civil War in the United States obliged many owners to sell off their vessels to foreigners, and the opening of the Suez Canal in 1869 was fatal. Sailing ships had difficulty negotiating the narrow confines of the canal in a contrary wind and were obliged to sail around the Cape. The newfangled steamships had no such difficulties, and their passage through the Suez cut off an enormous distance from the trip between Europe and China.

Only the *Cutty Sark* remains (figs. 3.14–3.15), her continued existence a testament to the enduring popularity of these great sailing ves-

Figure 3.14. The *Cutty Sark,* one of the last tea clippers to be built (in 1869) and the only survivor of a noble breed. Thanks to Jan van der Crabben for this image.

Figure 3.15. The *Cutty Sark* at night, showing some of her 11 miles of rigging. Image courtesy of The Cutty Sark Trust.

sels.* In her dry dock at Greenwich in the port of London, she has received over 15 million visitors. She may not have beaten her great rival *Thermopylae* on the high seas, but she has outlasted her by a century.

The Last Days of Sail

With hindsight we can see that, even without rough seas, unfavorable economics, and civil wars, steam engines alone would have killed off the clippers. James Watt made his major breakthrough in steam engine design the same year that HMS *Victory* was built; the age of steam locomotives coincided with the clipper revolution, and it was only a matter of time before steam engines replaced sails to drive ships. At the beginning of the nineteenth century, ships were made of wood and powered by wind; by the end they were built of iron and powered by steam. In

*Perhaps even she is gone. During the writing of this book, the *Cutty Sark* was extensively damaged by fire.

Figure 3.16. Model of a mid-nineteenth-century paddle steamer. A steam engine powered the paddle wheels athwartships, augmenting sail power. Thanks to Slava Petrov for this image.

between, there was an awkward period of hybrid technology, such as the paddle steamer (fig. 3.16).

Sailing ships were still being built to carry freight at the end of the nineteenth and beginning of the twentieth centuries, however; throughout the 1800s there was a proliferation of types of sailing ship, to be described here. The application of iron and steam was not uniformly adopted across the maritime world or for all applications. Sometimes sailing vessels were more appropriate or more economical, at least for a few decades longer.

When sail ruled the waters of the world, square-rigged ships dominated offshore travel, exploiting the prevailing winds on long ocean voyages. We saw in the last chapter how square-rigged ships sail well before the wind. Vessels rigged fore and aft sail better to windward (I will unpack the physical principles behind this fact in the next chapter) and allow greater control. They also require significantly fewer crew to operate—no need to go aloft. The apogee of full-rigged vessels, the clippers, carried both square and fore-and-aft sails, and by skillfully trimming the sails to exploit local wind conditions they sailed efficiently over perhaps as much as three-quarters of the compass. Clippers are still being built today for recreation or training—and because people like them (fig. 3.17).

By the end of the nineteenth century, purely fore-and-aft rigged ships

Figure 3.17. The modern Dutch clipper *Stad-Amsterdam*. In the nineteenth century, not only the Americans and British built clippers: the Dutch built over 100. Image courtesy of Bruno Girin / DHD Multimedia Gallery.

—schooners—were carrying cargo along the coasts of North and South America. These schooners were numerous and efficient; the gaff rigs required very few crew (perhaps 6 to 8) to manage. The fishing schooners that set out from eastern Canada to fish the Grand Banks are still fondly remembered in Nova Scotia. Some ships added a square-rigged topsail, useful in a light breeze or when maneuvering in harbor.

The following are brief descriptions of some of the types of nineteenth- and early-twentieth-century sailing ships. These summaries also provide the opportunity to recap the advantages and disadvantages of different sail plans.

- Fully rigged ship. At least three masts, all square-rigged.
- Brig. Two masts, both square-rigged (fig. 3.18). This is an efficient sail plan, and many brigs served as freighters right up until the end of the era of commercial sailing ships.

Figure 3.18. I'm calling this a brig since the two masts are (mostly) square-rigged. Despite the Caribbean location, I doubt the pirate flag. Thanks to Clayton and Fiona Lewis for this image.

~Hermaphrodite brig / brigantine. Two masts, with a square-rigged foremast and a fore-and-aft mainmast. The *Mary Celeste* was a brigantine.*

*The *Mary Celeste* was made famous in a fictionalized account by Sir Arthur Conan Doyle, creator of Sherlock Holmes. The true story of this mysterious ship is just as mysterious as Doyle's account. The American-registered vessel left New York for Genoa, Italy, on 7 November 1872 with an abstemious Bible-reading New England captain, his wife and two-year-old daughter, and eight Dutch crewmembers. None of them was ever seen again. She was carrying 1,700 barrels of raw alcohol. The ship was known to have passed the Azores on 25 November, but was found abandoned and drifting further east on 4 December. She was in good shape, there were no signs of violence, nothing had been stolen (though some of the barrels were empty), but there were 3 ft of water in the hold. The lifeboat and sextant were missing, though the crew's boots and pipes were found. A frayed rope trailed behind the abandoned ship. The British board of inquiry at Gibraltar could not decide what happened to the people on board, but starting with Sir Arthur, there have been many theories (some plausible and some outlandish) put forward.

Figure 3.19. A modern day barquentine. Thanks again to Clayton and Fiona Lewis for this image.

~Barque/bark. Three masts, with square-rigged foremast and mainmast and a fore-and-aft mizzenmast (see fig. 2.5). Like the brig, a very popular design; there were probably more barques built than all other square-rigged ships combined.

~Barquentine. Three masts, the foremast square with the other masts fore-and-aft rigged. See figure 3.19.

~Schooner. At least two masts, originally gaff-rigged, though nowadays any fore-and-aft sail may be substituted. (See fig. 3.9.) The mainmast—the tallest—is second from the front. Originating in seventeenth-century Holland, this vessel generated a number of variants (see next the few entries) and was a popular workhorse all along the coast of North America in the nineteenth century.

~Fishing schooner. Two masts, gaff-rigged, and typically with a main gaff topsail and fisherman's staysails.* The famous *Bluenose* (fig. 3.20) was a fishing schooner.

~Square-topsail schooner. The name describes the vessel; see figure 3.21. A popular and versatile combination of sails.

* Fisherman's staysails are four-cornered rather than triangular.

Figure 3.20. The much-celebrated Canadian fishing schooner *Bluenose,* which won many international races in the 1920s and 1930s. Note the schooner characteristics: mainmast second from the front, and both masts gaff-rigged. Image from Wikipedia.

~Four-masted schooner. Each mast carried less sail, to improve handling. These vessels could carry more cargo than the smaller schooners. Some New England versions had five, six, or even seven masts.

~Tern schooner. Three masts, gaff-rigged with triangular fore-and-aft topsails (fig. 3.21). In other words, a typical schooner sail plan, but with one extra mast. Tern schooners were mass-produced at the end of the nineteenth century.

~Ketch. Two masts, both fore-and-aft rigged, plus jib sail(s). See figure 3.22. The larger mast is forward (unlike a schooner), and the mizzen mast is forward of the rudder post. The mizzen sail is utilized to increase drive.

~Yawl. Like the ketch, though the mizzen mast is much smaller and aft of the rudder post; see figure 3.22. The mizzen sail is used for control rather than drive.

~Sloop. Any fore-and-aft rigged boat with a single mast. In the nineteenth century these boats were used as fishing vessels and for trade

Figure 3.21. *Top:* A square-topsail schooner. Image courtesy of Darillo. *Bottom:* A tern schooner. Image courtesy of Clayton and Fiona Lewis.

Figure 3.22. At left: A ketch. Image courtesy of Clayton and Fiona Lewis. *Top:* A yawl. Image from Wikipedia.

between the Caribbean islands and the United States. The sloop is particularly popular today because of the Bermuda sloop.

~Bermuda sloop. Triangular mainsail plus one jib. The most popular recreational sailing vessel in the world because it handles well enough with the wind and is optimum upwind. I will have a lot more to say about this vessel later.

~Catboat. A type of sloop with the mast set well forward. Gaff-rigged and very broad in the beam.

~Cutter. A sloop with a bowsprit and at least two headsails.

Some of these latter-day sail plans are illustrated in figure 3.23. Now that you are expert at picking out ships and boats from their sails, maybe you can identify the strange beast shown in figure 3.24.

I have now reached the end of my historical survey. From dugouts, we have progressed to the modern yacht. This pleasure craft takes on a

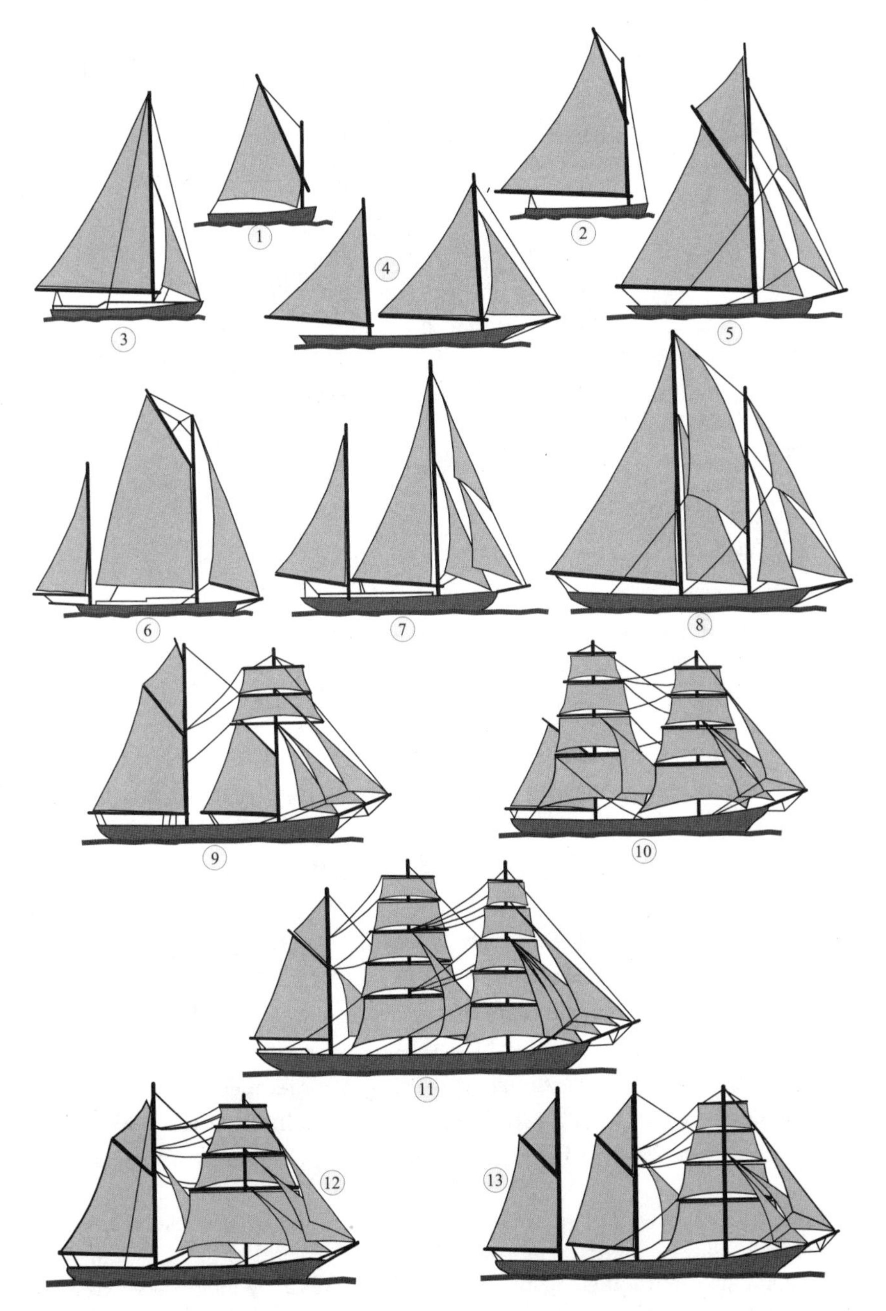

Figure 3.23. Types of sailing boats and ships: (1) sailing dinghy, (2) catboat, (3) knockabout, (4) Chesapeake Bay bugeye, (5) sloop, (6) yawl, (7) ketch, (8) schooner, (9) topsail schooner, (10) brig, (11) bark, (12) hermaphrodite brig, (13) barquentine. Adapted from *Virtue's Simplified Dictionary* (1948).

Figure 3.24. I am guessing that this vessel is a hermaphrodite brig (note the foremast yard). Any other suggestions? Thanks to Darillo for this image.

number of forms suggestive of earlier sailing vessels, as you have seen in some of the illustrations and as you can see in any large marina today.* The most popular sail plan today is undoubtedly the Bermuda-rigged sloop, and it is this vessel that I will analyze in the next chapter.

Let the cat out of the bag: *To disclose a secret.* Flogging with a knotted whip, the infamous cat-o'-nine-tails, was a brutal Royal Navy punishment administered by the bosun's mate for serious offenses. The "cat" was kept in a leather or cloth bag when not in use.

No room to swing a cat: *Insufficient space.* Crew members were obliged to witness floggings with a cat. These were performed on the open deck, because there was insufficient room below decks. Some sources say that

*Of course, technology has moved us along in many ways, so that the structure of sailing craft today is very different from that of a century ago. An obvious example is the material of construction: many boat hulls today are cavity-molded fiberglass—a material that did not exist when most hulls were painstakingly built of wood. However, the *physics* of sailing modern boats and older vessels is the same, as Scottie reminded Captain Kirk: "Ye cannae change the laws of physics, Captain."

for an unpopular flogging, the crew would crowd close to the bosun's mate so that he would not have enough room to swing the cat.

Over a barrel: *Placed in a predicament with no escape, an awkward position.* The harsh floggings administered by the Royal Navy in the past were carried out on deck. Prior to flogging, men were tied to a grating, whereas boys were tied over the barrel of a cannon.

Press into service: *Oblige to work.* In the eighteenth and early nineteenth centuries during times of war, Royal Navy "press gangs" scoured British ports for able-bodied merchant sailors between the ages of 18 and 55, even if they were not British. (In the U.S. equivalent "crimp gangs" on the West Coast would "shanghai" reluctant recruits.)

4

Analysis: Fore-and-Aft Boat Motion—Introducing Lift and Drag

Binge: *Heavy drinking session.* Originally a verb describing the rinsing out of a cask (in which liquids were stored aboard ship) prior to refilling.

Overbearing: *Bossy, arrogant.* A naval vessel would "bear down" on an enemy from upwind. When overbearing, she approached from directly upwind, thus robbing the enemy of wind power.

Overreach. *Extend too far.* A sailing ship that tacks for too long is said to overreach. To maintain the correct course, she must now increase her tack in the opposite direction to compensate.

Windfall. *An unexpected gift of money.* There are two claimed sources for this word, both of which are nautical and date back to the time of wooden sailing ships. (1) A sudden unexpected offshore wind that provides more leeway. (2) Trees felled by the wind in stands of timber that were otherwise reserved by the navy for building ships. (The landowner was free to dispose of such windfall trees as he wished.)

Aerodynamic Lift and Drag

The boat sketched in figure 4.1 may appear to be a square-rigger because the sail is oriented athwartships, but look at the (apparent) wind direction ($\underline{w}'$). The sail cuts the wind—it is a fore-and-aft sail—and the boat is not our old square-rigged friend *Snoozing Goose* but a more modern Bermuda-rigged sloop. Let us name her *Sparrowhawk*—more apt for a nippier and less somnambulant bird. For now the *Sparrowhawk* has her jib sail furled, not because this disposition best suits the weather she finds herself facing, but instead because we are not yet ready to analyze her with two sails deployed. First we must get up to speed—scientifically speaking—with how a single fore-and-aft sail provides drive, and only then introduce the jib sail and understand why Bermuda sloops are such a runaway success.

For those of you who are following the math in detail, endnote 1 is a quick reminder of my notation, which is the same as that used in chapter 2.[1] The main lesson to take from figure 4.1 is that the wind applies a force to a sail (or an airfoil) that is the sum of two forces: lift, $\underline{L}$, which is perpendicular to the apparent wind direction and drag, $\underline{D}$, which is in the same direction as apparent wind. The vector sum of these forces is the sail force—the total force that the wind exerts on the sail—and the component of sail force along the boat velocity direction is the drive, or boat force, $\underline{F}_{boat}$. Recall that in my momentum flux model, applied in chapter 2 to quantify the drive of a square-rigged boat, I approximated the sail as a flat plane and assumed that the sail force was perpendicular to this plane. We will see that this is a reasonable approximation in many circumstances but that it is not strictly true: the vector sum of forces $\underline{L} + \underline{D}$ is not necessarily perpendicular to the sail, and anyway, the sail is not a flat plane, as any sailor knows.

I will relegate the gory details of lift and drag theory to the appendix; here it is important only to understand that these two forces vary, as we will see, and that their sum is not always in the direction predicted by the momentum flux model. In this chapter we are stepping a few paces closer to a correct understanding of lift and drag, but I will not force-march you to this distant destination all at once. All we need to know for now are the expressions for lift and drag forces given in terms of lift and drag coefficients; from these we can explain much about fore-and-aft sail capabilities with the wind, across the wind, and to windward.

Lift and drag forces depend on apparent wind speed and other factors as follows:

$$L = \tfrac{1}{2}c_L \rho A w'^2 = \tfrac{1}{2}c_L F_{wind}, \; D = \tfrac{1}{2}c_D \rho A w'^2 = \tfrac{1}{2}c_D F_{wind}. \tag{4.1}$$

Here F_{wind} is the force exerted by the wind on the sails, as in chapter 2. The lift and drag coefficients, $c_{L,D}$, are the rugs under which an awful lot of hard physics is swept. In practice these coefficients vary with speed and other physical attributes of the wind, and also with sail orientation

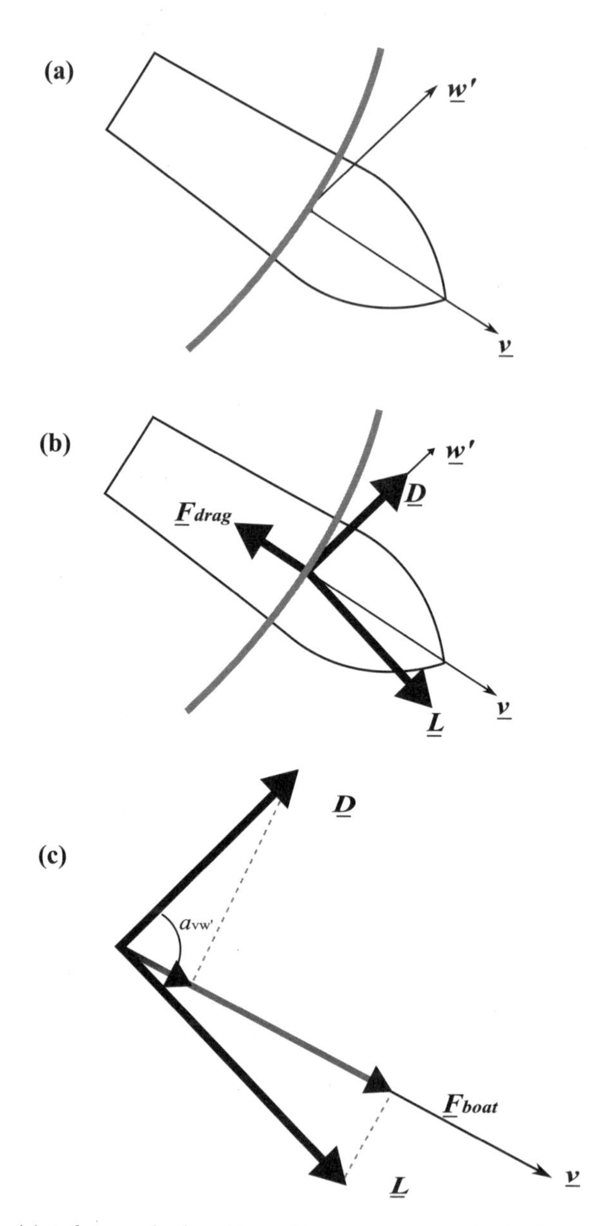

Figure 4.1. (a) A fore-and-aft sail (gray) in an apparent wind velocity $\underline{w}'$. The boat velocity is $\underline{v}$. (b) The aerodynamic lift force, $\underline{L}$, provided by the sail is directed perpendicular to the apparent wind direction. The aerodynamic drag force, $\underline{D}$, is in the same direction as $\underline{w}'$. Hydrodynamic drag is in the opposite direction to $\underline{v}$. (c) The drive $\underline{F}_{boat}$ provided to our boat by the wind force, via the sail, is the sum of lift and drag force components along the direction of $\underline{v}$.

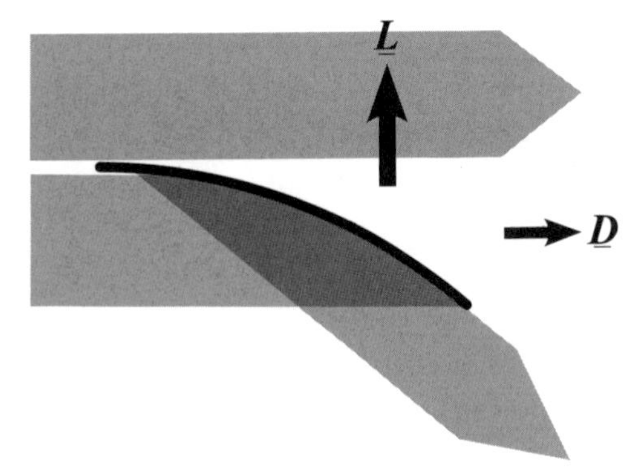

Figure 4.2. Air (gray arrows) flowing over and under a simple airfoil (black curve) generates lift and drag forces. Air pressure is reduced in back of the airfoil and is increased near the front (lower) surface. It is intuitively obvious that drag force will result from this airflow. We can get a handle on how a lift force might be generated by considering the pressure differences.

and shape.* One aspect of this coefficient variability will be allowed in the present chapter, enough to provide a reasonable account for observed fore-and-aft sail performance. So, my initial representation of lift and drag coefficients is simplified for ease of presentation.

It may seem strange to you that we physicists choose to consider two perpendicular components of the wind force, lift and drag, rather than simply consider the force as a single vector as we did in chapter 2. I think that I can provide you with an intuitive understanding of why we divide up the total force in this manner, without the need for math. In figure 4.2 you can see a simple airfoil that influences the flow of wind that passes over and under it.† The air passing beneath the airfoil is deflected downward, and the air passing over the top creates a pressure reduction (a partial vacuum) at the back of the airfoil. The partial vacuum occurs because air is blocked by the airfoil. Nature abhors a vacuum and seeks to fill the gap: air flows down and the airfoil is lifted up. Hence lift. Also, it is easy to imagine how deflected air beneath the wing causes a pressure increase in this region, which you might consider, à la momentum flux, to push the airfoil upwards. This is why the lift force is given its name: for an airfoil that describes an airplane wing, the airplane is indeed lifted

*The shape of a sail is three-dimensional, but even if simplified to two dimensions (as suggested in fig. 4.1) the lift and drag coefficients change as the curvature and depth of the sail—the airfoil—change. This variability is well-known from airfoil theory; you can read more about it in the appendix and in the technical references of the bibliography.

† A sail acts like an airplane wing but is more complicated because it is not rigid. This complication is ignored for now and both wing and sail are regarded as simple airfoils.

upward. For a sail, however, the "lift" is sideways, perpendicular to the direction of air flow.

This simple, intuitive view of lift suggests (correctly) that lift force is composed of two contributions: pressure difference and momentum flux. We will see in the appendix that a more complete picture of aerodynamic lift finds the same two contributions.

Drag is equally intuitive, and again the simple momentum flux approach provides a motivation. Clearly the air that impinges on the undersurface of the airfoil (the weather side of the sail) in figure 4.2 will impart a force that tends to carry the airfoil with it. Put another way, airfoil movement through the air will be resisted. This is drag—in part. Physics is more complicated than this simple explanation suggests: in fact, the reduced pressure on the upper side of the airfoil also contributes to drag. Vortices form at the leading and trailing edges of the airfoil and are shed, carrying energy away with them. More details are provided in the appendix, but for now all we need is an appreciation that an airfoil or sail is subjected to two forces at right angles to each other. The relative magnitudes of these forces—the much-discussed L/D ratio—depends critically on airfoil orientation to the airflow. This orientation, specified by the airfoil angle to the airflow, is called the *angle of attack*. The angle of attack is shown in figure 4.3, where it is defined using the airfoil chordline.

The dependence of lift and drag on the angle of attack is crucial in determining the effectiveness of an airfoil, be it an airplane wing or a fore-and-aft sail. It is interesting and instructive to consider our old momentum flux approach of chapter 2 from the point of view of lift and drag. This can be done—the two approaches are not mutually exclusive, as is sometimes thought—and is achieved as shown in figure 4.3. We take the chordline of our airfoil to represent the momentum flux sail (which is assumed to be flat). We calculate the wind force, $\underline{F}_{wind}$, and then project it onto the sail direction to obtain the momentum flux sail force, $\underline{F}_{sail}$ (fig. 2.3). Now, if the momentum flux approach is the same as the modern aerodynamic lift and drag approach, then it must be the case that $\underline{F}_{sail} = \underline{L} + \underline{D}$. In words: if the momentum flux approach is strictly correct, then the momentum flux sail force must equal the aerodynamic lift force plus drag force combined, as shown in figure 4.3. We know that in the real world this equality does not hold because the momentum flux

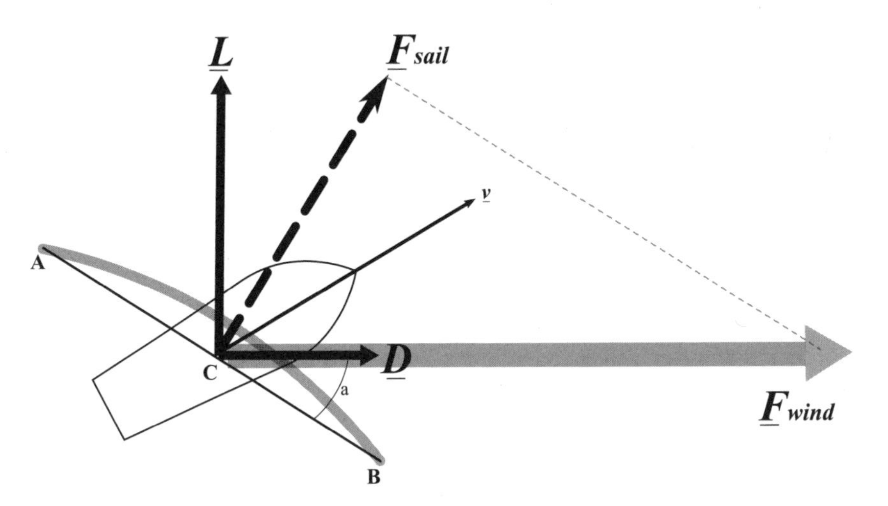

Figure 4.3. A fore-and-aft sail (curved gray line), here regarded as a simple two-dimensional airfoil, derives a sail force, $\underline{F}_{sail}$, from the wind force, $\underline{F}_{wind}$. This force can be broken down into lift and drag components. If the sail force is perpendicular to the chordline ACB, our intuitive momentum flux analysis produces the same results as the more accurate lift and drag analysis. The two approaches broadly agree for downwind sailing, as here; but for sailing into the wind we must use lift and drag analysis. In this approach, the forces of lift and drag depend critically upon angle of attack, *a*.

approach is not exact, but please bear with me because the comparison is a fruitful one. From figure 4.3 we can show that, if lift and drag coefficients depend on angle of attack as follows, then the momentum flux approach is the same as the lift and drag approach:[2]

$$c_L = 2\sin(a)\cos(a),\ c_D = 2\sin^2(a). \tag{4.2}$$

Here a is angle of attack.[3] These equations are plotted in figure 4.4 for different angles of attack. Real lift and drag coefficients depend on angle of attack differently; the plots (I will call them "lift and drag curves") vary with airfoil shape, air density, and other factors discussed in the appendix, as well as on angle of attack. So, in reality we obtain a number of different lift and drag curves, depending on the airfoil design. One such curve, more representative of real lift and drag characteristics than equation (4.2), is also plotted in figure 4.4.

Now we can understand why the momentum flux approach worked well enough for square-rigged vessels: the momentum flux curve is quite

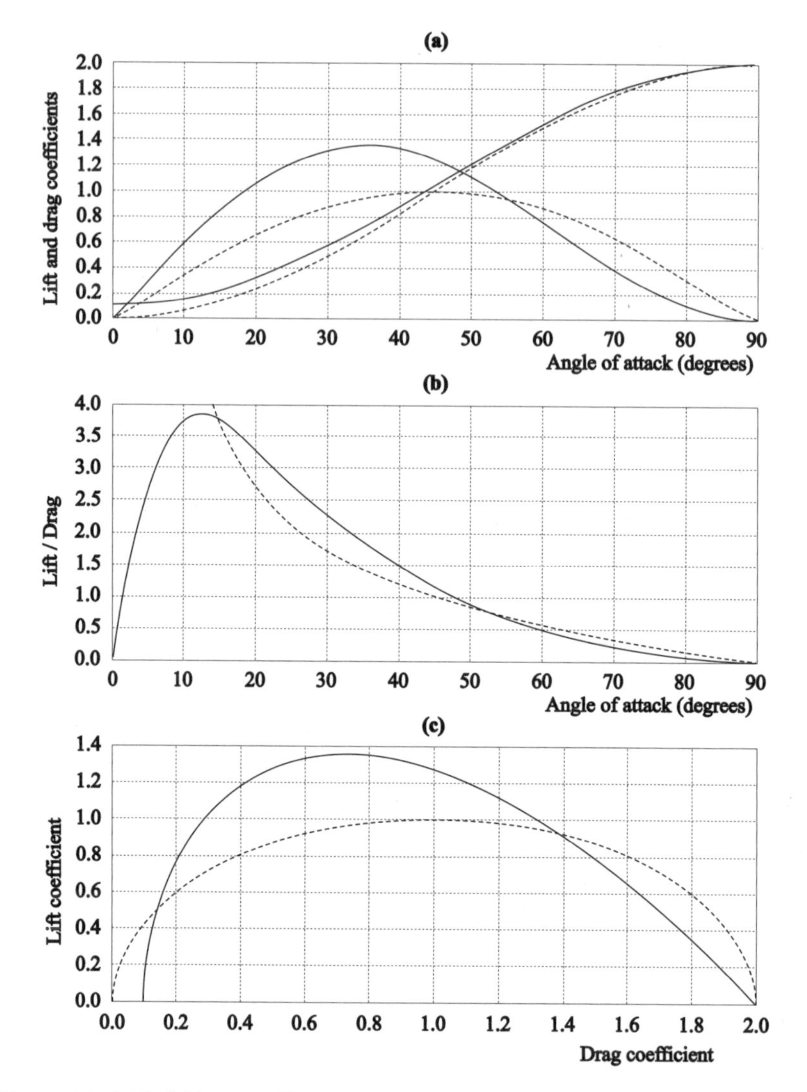

Figure 4.4. (a) Lift/drag coefficients CL and CD vs. angle of attack (AoA). The drag coefficients are a maximum for angle of attack = 90°. Dashed lines correspond to the momentum flux curves (eq. [4.2]), whereas solid lines correspond to more realistic lift and drag coefficients. (b) Lift-to-drag ratio vs. AoA. A peak value of 3 or 4, at an angle of attack of between 10° and 15°, is typical of realistic lift/drag behavior. The dashed line again corresponds to momentum transfer values—unrealistic for small AoA values. (c) Lift vs. drag. The exact shape of this curve for different sail plans depends upon many factors and can vary markedly. The solid line curve shown here is representative. Note that for AoA = 0° (extreme left) drag is zero for the momentum flux curve (dashed line), since the sail is edge-on to the wind. In reality, some drag remains even for edge-on sails.

similar to actual curves but differs in detail. The differences at low angles of attack turn out to be important (we will see why very shortly), but at a large angle of attack they are not important. Square sails running with the wind—which is what they do best—are set to catch as much wind as possible, so that the angle of attack is large (90°). When one is sailing closer and closer to the wind, it is necessary to trim the sails so that the angle of attack is smaller and smaller, and this is what fore-and-aft sails do best. So, the momentum flux approach models actual lift and drag quite adequately for downwind sailing but is less accurate for sailing across the wind and still less accurate for sailing to windward. From figure 4.1 or figure 4.4 you can see that for sailing downwind the drive is provided principally by drag force, whereas in sailing close to the wind the drive is mostly provided by lift. Our momentum flux approach is closely akin to drag because the drag force, *D*, acts in the same direction as the wind force, F_{wind}. This is why the difference between the momentum flux curve of figure 4.4 and more realistic lift and drag curves is unimportant at a large angle of attack: momentum flux is largely equivalent to drag,* and drag dominates at large angles of attack. Momentum flux does permit sailing to windward (that is, it gives rise to a kind of lift force) by projecting F_{wind} onto the sail direction, as shown in figure 2.3, but herein is a flawed assumption. The momentum flux approach *requires* that the sail force, F_{sail}, resulting from this projection be perpendicular to the chordline AB of figure 4.3. In reality this is not the case for small angles of attack: i.e., $\underline{L} + \underline{D}$ is not necessarily perpendicular to the sail, as stated earlier.

So, for sailing close to the wind the momentum flux approach of chapter 2 is no longer adequate; we bid it adieu and move on to examine what our modern views about lift and drag add to our ability to sail upwind.

*A purist might squirm at this strong assertion because the causes of drag are more complex than might be supposed by considering momentum flux alone. (For example, momentum flux says nothing about vortex formation and shedding, which are crucial to a detailed understanding of drag aerodynamics.) However, my aim is to convey general ideas and principles, not detailed physical technicalities. I hope that most purists will understand my approach. Some won't: I would like to transport them to the topgallant yard of a clipper barrelling homewards in a strong following wind, and ask them to reconsider.

Equilibrium Speed on Hard Water and Downwind

Despite the foregoing discussion, equation (2.1) of chapter 2 still applies:

$$F = F_{boat} - mbv^2. \tag{4.3}$$

Here, however, the form taken by the boat force, F_{boat}, is different. It is now determined by projecting the lift and drag forces along *Sparrowhawk*'s velocity direction. As in chapter 2, I will assume that our boat makes no leeway; her centerboard or keel resists lateral movement. Again, this is an idealization or approximation that leads to insight at the expense of technical rigor. Even so, the resultant differential equation is painful and generally can be solved only by number-crunching in a computer.[4] Certain special cases, however, can be solved by resorting to nothing more complex than high-school algebra, and so I will begin with these.

Suppose that the hydrodynamic friction is very small—perhaps *Sparrowhawk* is sailing on a lake of liquid helium[5] or, more realistically, is an iceboat. For this case the math can be simplified enough for me to write down the expression for boat equilibrium speed—the fastest she can go in her chosen direction, given enough time.[6] It is

$$v_{eq} = w\left[\frac{c_L}{c_D}\sin(a_{vw}) + \cos(a_{vw})\right] \tag{4.4}$$

So the math tells us that equilibrium speed for an iceboat increases with wind speed—no surprises there—and that it depends on point of sail and on L/D ratio. If the adrenalin-addicted pilot of the iceboat* sets his sail at a constant angle of attack,† then the lift and drag coefficients are unchanging as his boat picks up speed. In this case we can calculate the best point of sail and the maximum equilibrium speed attainable along that direction.[7] The maximum equilibrium speed is $v_{eq} = w\sqrt{1+(c_L/c_D)^2}$ and occurs when the iceboat is heading at an angle to the wind, a_{vw}, given by $\tan(a_{vw}) = c_L/c_D$. Realistic figures for maximum attainable aerodynamic L/D ratios depend on design, but typically are 3:1 or 4:1 (see fig. 4.4), corresponding to heading angle, a_{vw}, of a little over 70°. So our analysis tells us that if Captain Intrepid sends his iceboat on a broad reach

*Is there any other kind?

†As the iceboat picks up speed, the apparent wind direction will change and so, to maintain constant angle of attack, the pilot must continuously adjust his sail.

at an angle of 70°–75° to the downwind direction, he may attain speeds of three or four times the wind speed. Iceboats can indeed achieve such astounding results.* It is an age-old conundrum among landlubbers ("how can a boat sail faster than the wind?") but is familiar to the nautically inclined, though the extremes achieved by "hard-water sailors" are exceptional. So our analysis of low-friction sailing produces answers that accord with observation: the highest speeds, of several times the wind speed, are achieved sailing across the wind. In the absence of hydrodynamic drag, the limits to speed are determined by the aerodynamic lift-to-drag ratio.

Encouraged by the credible results of our analysis, we look for another special case that is easy to investigate. We find one: running with the wind. When the *Sparrowhawk* is moving in the same direction as the wind, we can solve the equation of motion exactly with pencil and paper, without the need for computers. The result is an equilibrium speed of $v_{eq} = w/(1 + \sqrt{2\beta / c_D})$. This is the same as the expression for a square-rigged boat heading directly downwind, except that the parameter β has been replaced by $2\beta / c_D$. Now recall that, as indicated in figure 4.4, we expect the drag coefficient c_D to be about equal to 2 for a boat running with the wind because in these circumstances the sails will be set at an angle of attack of $a = 90°$ to maximize the drag. (The lift force is nowhere when the boat is running downwind; drive comes from drag.) So the performance of a fore-and-aft rigged boat running with the wind is comparable to that of a square-rigger. Perhaps the performance is not quite so good, because fore-and-aft sails are smaller (with less effective area, A) than those of a fully rigged ship. Smaller A means larger β means smaller equilibrium speed; square-riggers are better downwind sailors because they can pile on more canvas. Size matters when running with the wind, it seems.

Beating to Windward with a Fore-and-Aft Rig

When heading upwind, it is a different matter altogether. We know this from experience. Is the renowned superiority of fore-and-aft rigged boats vindicated by physics? If not, there is something wrong with our

*The official world record for iceboat speed was set back in 1938 by John D. Buckstaff on Lake Winnebago, Wisconsin. His craft *Debutante* glided over the ice at 147 mph. According to *Encyclopaedia Britannica,* iceboats can achieve speeds "at least four times" the wind speed when sailing across the wind.

physics because any Bermuda sloop is unquestionably much better than any square-rigger when pointing into the wind. But how much better, and why?

The "why" we can answer already, from what we know of square and fore-and-aft rigging. Square sails are hung from yards and may be trimmed, via bowlines, to catch a following wind or to slant the sail at an angle, enabling *Snoozing Goose* to sail on a beam reach or a few points to windward. We know that the angle of attack must become smaller and smaller as the *Goose* points closer and closer to the wind,[8] and this causes the sail to luff. The fore-and-aft rigged *Sparrowhawk* performs better because the luff of her mainsail is stiffened by the mast. Were she gaff- or lateen-rigged instead of Bermuda, the same would apply: the edge of the sail presented to the wind is stiffened by the angled spar. This stiffening permits the sail to cut the wind rather than luff up uselessly. For staysails, stiffening is provided by the stays rather than by a mast or spar, but the same result obtains. Fore-and-aft sails perform better to windward than do square sails because they are better designed to hold their shape when close-hauled.

To see how much better the Bermuda rig performs, I must bite the bullet and tackle the full equation of motion, which some of you have already seen in note 4 but which I have avoided in the text until now. In fact, by confining my attention to the equilibrium speed I avoid calculus and am left with algebra, which is easier. The results are shown in figure 4.5. Here, the approximate equilibrium angle[9] for a Bermuda-rigged boat (characterized by the realistic lift and drag curve of fig. 4.4) is plotted for different wind directions, from running with the wind to close-hauling. According to my analysis, *Sparrowhawk* can sail to within about 20° of the wind, though she will make no headway. At 30° she is making pretty good headway, moving upwind at 10% of the true wind speed. Recall that *Mir* performs exceptionally for a predominantly square-rigged ship—and even she can sail no closer than 37° (see fig. 2.10). *Sparrowhawk* can make headway at around 15% of wind speed when four points (45°) off the wind, which is the limit for close hauling for many square-riggers. Note also that the maximum speed of *Sparrowhawk* can get up to about 30% of wind speed—pretty nippy compared to a square-rigger, though a racing yacht will do better and may exceed wind speed. The downwind performance is comparable to that of *Snoozing Goose*—perhaps not so good, but we have not yet unfurled *Sparrowhawk*'s headsails.

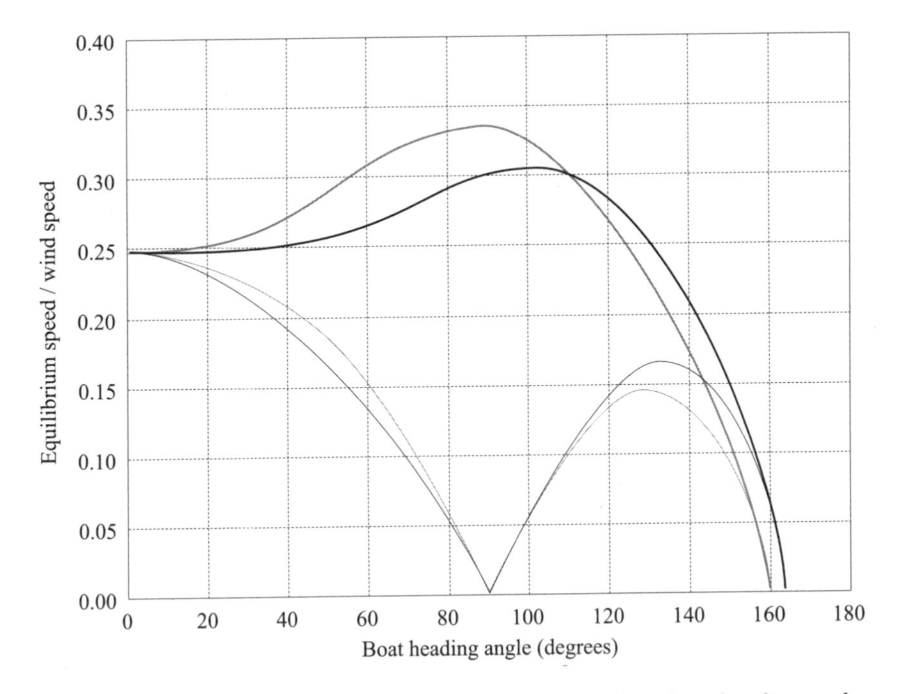

Figure 4.5. Equilibrium speed, as a fraction of wind speed, vs. boat heading angle a_{vw}. For $a_{vw} = 0°$ the boat is heading downwind, and for $a_{vw} = 180°$ she is heading straight upwind. Bold curves show boat speeds for two different helmsmen, who trim their Bermuda rigs a little differently. Maximum speed occurs on a beam reach. Thin lines show the corresponding component of boat speeds upwind and downwind. Note that our fore-and-aft rig can sail to within about 20° of the wind. Downwind performance for this boat is not significantly different from that of a square-rigger.

In chapter 2 we saw that a sailboat can create its own wind but that square-riggers were not set up to do that. We can expect that fore-and-aft rigged boats will do better in this respect, and the physics backs us up. Calculating the rate of change of acceleration with time[10] from the fore-and-aft equation of motion, we find the results shown in figure 4.6. Here the critical parameter for upwind performance is the L/D ratio. Depending on how the boat is handled during close hauling, *Sparrowhawk* and her fore-and-aft kin can indeed create their own wind, over a limited range of heading angles. The calculation leading to the plots shown in figure 4.6 serves to show not only the superior upwind performance of fore-and-aft vessels but also the significance of the helmsman. His skill and experience are utilized to set the sail angle of attack (determining

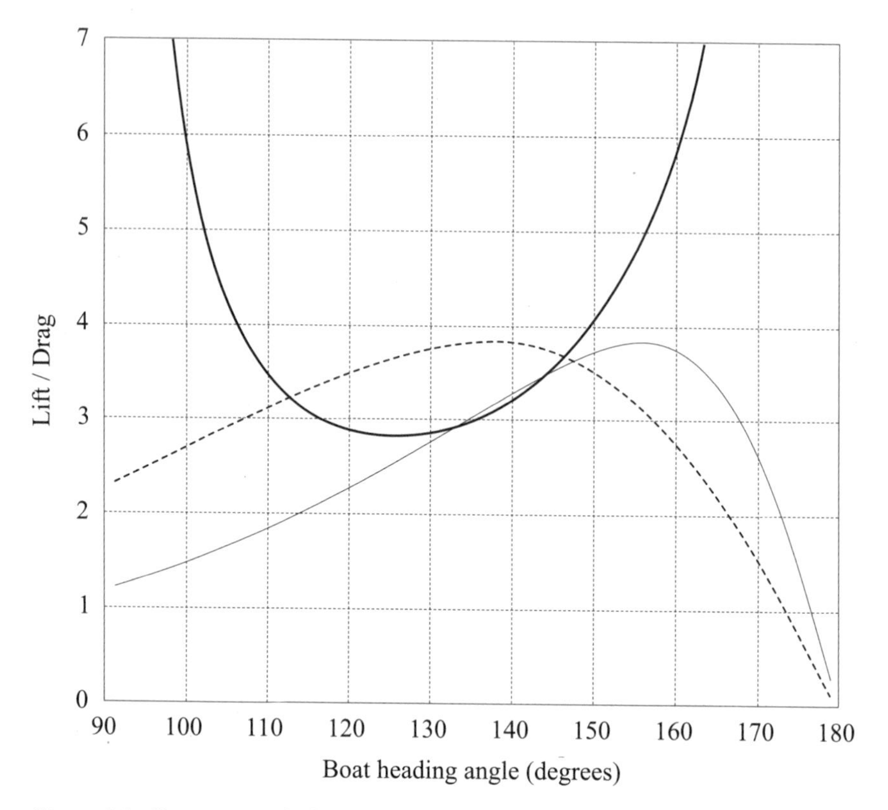

Figure 4.6. Creating wind: If lift/drag ratio L/D is above the bold line for a given heading angle a_{vw}, *Sparrowhawk* creates her own wind and her initial acceleration increases. Otherwise she does not, and her initial acceleration slows down. The thin line applies for the sail trim maintained by helmsman H1, whereas the dashed line is that of helmsman H2 (see note 9 for details). The point is that different sail trimming leads to different upwind capabilities; here, H2 enables *Sparrowhawk* to create her own wind over a wider range of heading directions than H1.

where on the lift and drag curve the sail is operating) and also the sail shape (which influences the shape of the lift and drag curve). Two helmsmen operating the same vessel may form very different opinions about how much she can create her own wind.

The Story So Far

At this point I can briefly summarize the results of the analysis of sailing performance for square-rigged and Bermuda-rigged vessels heading downwind, across the wind, or upwind. If we compare the plots in figures

2.6 and 4.5 for downwind sailing or sailing on a broad reach, we see little performance difference when the two boats have the same sail area. Downwind, the lift and drag approach yields results very similar to those obtained using the simpler and more intuitive momentum flux approach; *Snoozing Goose* probably beats *Sparrowhawk* by virtue of her greater sail area. It is clear from the physics of lift and drag forces that fore-and-aft sails generate more drive from the wind than do square sails when sailing to windward, or even on a beam reach. Our simple momentum flux analysis in the upwind direction is probably not very accurate because this approach neglects some physics which matter when close hauling. Anyway, the construction of square sails is optimized for downwind sailing, and even neglecting the effects of luffing when the boat is close-hauled, we found that *Snoozing Goose* could do no better than 10% of wind speed at 45° to the wind. In practice, we can expect she will not perform so well. *Sparrowhawk,* on the other hand, can comfortably exceed 10% of wind speed when reaching and hauling between about 70° and 30° of the wind. Her sails are easier to manage and so require less crew; they can be braced over a much wider range of angles so that a skillful helmsman can set the angle of attack to maximize lift when lift is needed and drag when drag is needed.

The same straightforward lift and drag analysis based on applied forces and the corresponding differential equations showed that *Snoozing Goose* probably cannot generate her own wind—that is to say, she cannot use her own speed to increase the driving force. *Sparrowhawk,* on the other hand, can do so, to an extent that depends on her helmsman. The bottom line of this analysis confirms our experience and provides a theoretical underpinning for why it should be so: square-riggers are downwind specialists, and fore-and-aft vessels are better all-rounders.

One Sail, Two Sails, Foresails

Thus far I have restricted *Sparrowhawk* to deploying one sail. Most modern boats have at least one foresail (a.k.a. a headsail)—a jib or a spinnaker. Here I will examine the physics underlying these extra sails and the benefits of a combination of two sails.

First, though, an exercise in elementary geometry to provide one reason that triangular fore-and-aft sails are more effective than square sails of the same area. Consider figure 4.7; the square sail has the same

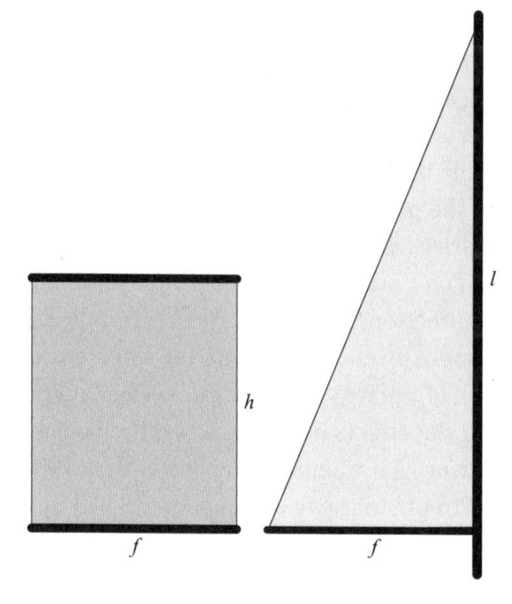

Figure 4.7. Triangular sails are more rigid than square sails, since only one edge is free. They also are more aerodynamically efficient because the effective sail area is greater. For the two sails shown the triangular sail has an aerodynamic area four times larger than that of the square sail, even though the physical area is the same (because $l = 2h$).

physical area of canvas as the triangular sail, and yet the triangular sail has greater effective aerodynamic area. Why so? Recall from chapter 2 that because air is a fluid it flows around obstacles such as sails, and the streamlines are distorted even for air that does not physically contact the sail. The significance is that the area of air deflected by the sail is $A = \pi x^2/4$, where x is the larger dimension of sail width and height. So, for the square sail of figure 4.7 the physical area is fh, whereas the effective aerodynamic area is $\pi h^2/4$, which is not necessarily the same.* The actual area of the triangular sail is smaller if l is less than $2h$, but the effective area is greater if l is bigger than h. So, even if triangular sails are physically smaller, if they are tall they catch more of the wind and so represent more efficient sails. (In fig. 4.7, $l = 2h$ and so aerodynamic area is πh^2.) This consequence of geometry is well known to sail designers and is seen every day in the very tall triangular sails of modern fore-and-aft boats (see fig. 4.8). There is more to the physics of triangular sails than I have indicated here, and so we will return to this subject later.

Another factor that both helps and hinders sailing is the wind speed profile. Because of friction with the ocean surface, and because air is a

*Here we are regarding the square sail as an airfoil, albeit one that is designed to produce drag and not lift.

viscous fluid, air speed at the surface is zero but increases with altitude.[11] This phenomenon is very similar to fluid flow in a pipe—say, blood flow through veins and arteries—whereby friction with the pipe walls slows flow at the edges but not in the middle, leading to a radial speed distribution. In the case of wind, the increase in air-flow speed with altitude leads to the phenomenon of twist, well known to sailors. As you can see from the vector diagram of figure 4.9, the increase in true wind speed with

Figure 4.8. A tall fore-and-aft sail has larger aerodynamic area than a squat, square sail of the same physical area. Birds' wings are long and thin for the same reason, and also generate a lot of lift for their size. Note the close resemblance of the sail to the seagull's wing. Thanks to Simona Manca for this image.

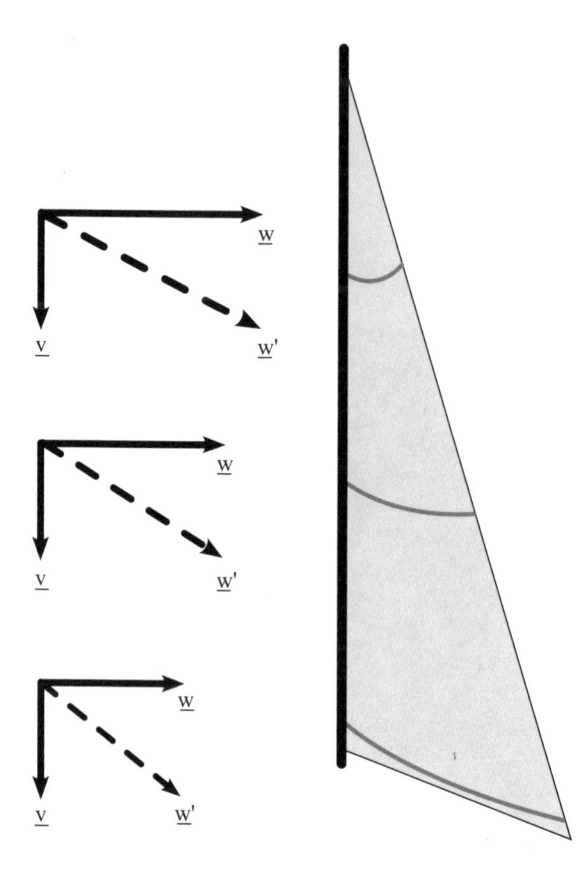

Figure 4.9. Wind gradient influences apparent wind direction as well as magnitude, here illustrated for a beam-reach point of sail. The triangular fore-and-aft sail angle of attack varies with height and, because sails are not rigid, so does the depth (curvature) of the sail.

distance up the main mast means that both apparent wind speed and direction change with distance up the mast. This is twist. A helmsman may set his angle of attack just right for the foot of the tall triangular sail shown in figure 4.9, but the angle of attack at the head may be quite different.*

Experienced helmsmen take this into account. They can influence both sail shape and angle of attack via two controls: the mainsheet and the vang. The sail fullness (or curvature, illustrated in fig. 4.9) changes with angle of attack and wind speed because sails are not rigid airfoils, and so the lift of a sail (per square foot) changes with height up the mast. Also, of course, the area of a triangular sail that is adjacent to any given

*Sails can be "twisted" up the mast to compensate, i.e., to maintain a constant angle of attack at all heights. This may be one reason that tall ships have many sails on one mast, as shown in figs. 2.5 and 2.8.

inch or foot of mast decreases with height. All of these factors make the aerodynamics of sails—particularly the tall fore-and-aft sails—very complicated indeed and difficult to simulate accurately, even with high-powered computers.

Those of you who are heavily into the kind of number-crunching simulations that must be performed to convincingly simulate the effects of twist on a modern fore-and-aft sail may consult some of the references provided in the bibliography; these are as close as you are going to get to this type of analysis in my book. Nothing against high-octane turbo-powered computer simulations, please understand—they are important for all yacht designers and essential for racing yacht design—but they go against my philosophy here. I merely point out the existence of twist, noting its simple physical origins, and say that the consequences for yacht motion prediction are too complex for a handy straightforward calculation to capture accurately.

For you to gain a deeper understanding of the value of triangular sails when sailing to windward, I need to say something about vortex formation and shedding. This subject is too complex for this chapter and so is postponed until the appendix. One consequence of the wind speed profile is quite clear, however. The fact that wind speed is higher for topsails than for mainsails means that topsails are particularly effective for a downwind point of sail.* Topsail deployment is shown clearly in figure 4.10, where a square topsail and a triangular jib both provide drive for boats that are running with the wind. The topsail will be more effective because it is square (best for running downwind) and because it is higher up (greater wind speed). This photo illustrates the role that square-rigging can still play in fore-and-aft boats and ships, just as fore-and-aft staysails helped the basically square-rigged clippers. The combination of both types of sails provides for effective drive to wind and to weather.

For racing with the wind, the helmsman of a Bermuda-rigged yacht such as the one on the left in figure 4.10 might even let out the mainsail on the other side to the jib. Such a deployment is known as "wing-on-wing." This configuration approximates a square sail in shape, but because of the fore-and-aft rigging it is notoriously unstable. A slight imbalance in the drive applied to the two sails will cause the boat to

*Topsails also induce heeling through their large torque, but this is not important when running with the wind. Torque is the subject of the next chapter.

broach—i.e., to suddenly veer off course—which in turn may lead to the mainsail or jib flying dangerously across to the other side as it catches the changing apparent wind. The more sensible and sedate pace of the yacht in figure 4.10 is safer, though the helmsman must steer to compensate for veering off course due to the asymmetric wind load. The helmsman of the square-topsail schooner has no such problems because his topsails are symmetric and induce no shift in heading while running downwind. As I have mentioned several times already, for sailing downwind square sails win; fore-and-aft come into their element if you want to come back home.

Figure 4.11 shows a Bermuda sloop running with the wind in a safe and sedate fashion: she is being pulled along by her jib instead of pushed along by her mainsail. There is little danger of broaching in this case. If we think of a yacht powered by her mainsail as analogous to a rear-wheel-drive automobile, then the sloop of figure 4.11 is like a front-wheel-drive car. Front-wheel-drive vehicles are also pulled along and

Figure 4.10. Old and new ways of running with the wind. The old schooner on the right has gaff-rigged fore and main sails, plus a jib topsail, a jib, and a staysail (all fore-and-aft). The square-rigged topsails aid downwind movement. The modern Bermuda sloop on the left uses a jib for the same purpose. Thanks to Darillo for this image.

Figure 4.11. A Bermuda sloop running under jib alone, with mainsail furled. Thanks to Clayton and Fiona Lewis for this image.

exhibit less inclination to skid out of control. On the other hand, they lack the power provided by the rear wheels.

Fore-and-aft yachts can proceed downwind quite well at moderate speeds but experience problems when trying to run with the wind at speed. Spinnaker sails illustrate this basic point in a slightly different way. We have seen how square sails are deployed to increase drag force rather than to generate lift, whereas triangular fore-and-aft sails generate high L/D ratios, in particular if they have a high aspect ratio. The racing yachts of today, such those shown in figure 4.12, have very tall and thin triangular mainsails to provide an effective drive on all points of sail

Figure 4.12. Top: This Finn class racing yacht flies a high-aspect-ratio triangular mainsail to provide very effective lift and a much larger spinnaker to provide very effective drag. Symmetric spinnakers such as this one (image from Wikipedia) are usually deployed only for downwind movement, because (*bottom*) even a little side wind causes significant heeling (image courtesy of Clayton and Fiona Lewis).

from a broad reach to close hauling, but the very fact that they are optimized for high L/D ratios means that they do not perform well downwind, where drag provides the drive. Hence, the modern equivalent of a square sail, the spinnaker, is deployed. This foresail, called a "kite" or a "chute" because of the way it is used, is often twice the size of the mainsail. Its existence is elegant testimony to the ineffectiveness of fore-and-aft sails when running with the wind. Spinnaker and mainsail in combination make for fast sailing to wind and weather—if the crew works hard enough.*

The Slot Effect

The effect of sails in combination is not simply additive. Two sails can be twice as effective as one sail or less than twice as effective. The combination of jib and mainsail in a Bermuda sloop is very common because these two sails form a particularly beneficial symbiosis. The governing aerodynamics of this happy union goes by the name of the *slot effect*, about which much—too much—has been written over the years. By now the warning bells should already have rung in your head: as soon as the word "aerodynamics" is mentioned, you should anticipate conflicting claims and arguments over the physics of the slot effect. As always, I will banish detailed explanations, and also a brief summary of the common misconceptions, to the appendix. It is appropriate here, nevertheless, to at least state the cause and effect of "the slot" because it is an important part of sloop performance.

Airplane designers know all about the slot effect. The flaps on an aircraft wing, both at the front and the back, are control surfaces that permit the pilot to change the way that air flows over the wing. Staggered wings on an old biplane also influence each other. Way back in the 1930s Prandtl was able to write that "the total induced drag of a biplane is smaller than that of a monoplane of the same span and of the same lift."

*Symmetric spinnakers can be difficult to use and are hard work because they are flown only when the boat is running downwind and must be doused for other points of sail. The spinnaker of fig. 4.12 is symmetric (note the spinnaker pole that separates the mast from the spinnaker weather-side *clew*). There are also asymmetric spinnakers that can be deployed to windward. For downwind sailing there are also *genoa* foresails ("jennies"), which are large jib sails, and the so-called *gennakers*, which are genoa-spinnaker hybrids.

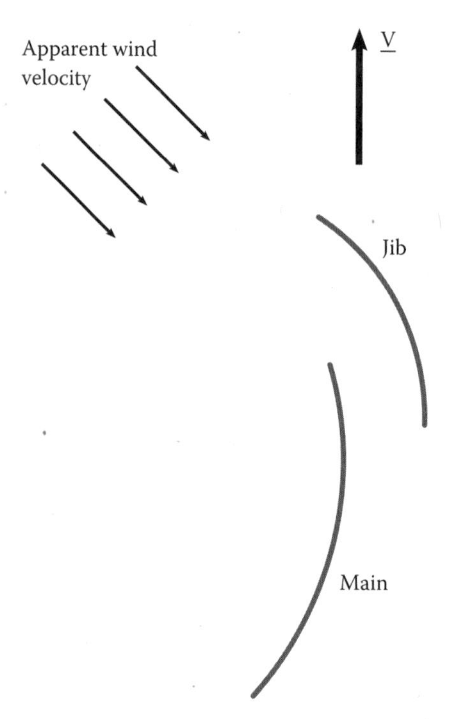

Figure 4.13. The slot effect is well known to airplane designers but has been understood in the context of sailing only within the last 40 years. On a windward point of sail, the jib influences, and is influenced by, the mainsail. The jib redirects the wind flow so that mainsail drag is reduced. The mainsail increases the jib's lift.

This fact may seem surprising, and certainly to someone schooled only in the Newtonian momentum flux ideas: the frontal area of biplane wings is greater than that of a single monoplane wing, so shouldn't the drag be greater? Well, no—that's aerodynamics for you. Explanations later. The important point is that drag is reduced, and the same thing applies to sails.

Jib and mainsail combine in the same way as the staggered wings of a biplane. When beating to weather, the combination is a very efficient sail plan. The typical orientation of jib and mainsail, and of apparent wind and heading directions, is shown in figure 4.13. The jib redirects wind on the lee side of the mainsail (jib and main form a "slot" for the air passing between) so as to reduce mainsail drag. Another effect, surprising and perhaps counterintuitive, is that the mainsail influences jib performance. How can this be, given that the mainsail is mostly downwind of the jib? All explanation deferred.

All that the practical helmsman needs to know is that the mainsail's presence makes the jib a more effective fore-and-aft sail. If the jib has a mainsail behind it, the jib provides more lift than it would in isolation. When racing on a close haul, the helmsman "shifts through the gears": he

Figure 4.14. Aerodynamics in action. The airplane wing provides vertical lift; the boats' triangular mainsails provide horizontal lift, whereas the foresails provide drag for downwind movement. What this picture does not reveal is the physics that applies underwater. Thanks to Simona Manca for this image.

repeatedly trims the jib and main as his yacht picks up speed. The jib is adjusted first because it influences how the mainsail should be trimmed, due to the slot effect.

Drive On

This chapter has been concerned mostly with lift and drag viewed as two perpendicular forces that can be exploited in combination by a well-chosen sail plan to provide drive for a sailing yacht. We associate lift with modern fore-and-aft sails and drag with square sails or big, ballooning ("highly cambered" is the more correct technical term) foresails such as spinnakers or genoas. I have shown that the manner in which lift and drag coefficients depend on angle of attack accounts for much of the observed sailing characteristics of fore-and-aft sails. (I have not shown you *why* lift and drag coefficients depend on angle of attack in the way that they do. Some of these reasons will be addressed in the appendix.)

The combination of two or more sails increases the ability of sailing yachts to fine-tune their response to the wind, extracting maximum drive over a wide range of points of sail. Sail combinations may be adjusted to adapt to changing wind direction or speed, resulting in yachts that can sail the oceans of the world in all directions and in all but the worst weather conditions.

This summary concludes my elucidation of the physical principles that underlie the driving force provided by sails. The remaining chapters deal with other aspects of sailing physics. Figure 4.14 is as good an image as any for me to close my discussion of drive; I hope that from now on when you look at a sailing vessel, be it a modern fore-and-aft yacht or a fully-rigged tall ship, you can appreciate a little better the harnessing of physics to provide useful—and enjoyable—sail power.

Between the devil and the deep blue sea: *Presented with a choice of equally unpleasant options.* The "devil" was the longest seam of a wooden ship in the Age of Sail, running from stem to stern along the length of the beam supporting the gun deck. To seal this seam at sea, a sailor was suspended over the side in a bosun's chair.

Cut of his jib: *Appearance.* The "cut" of a sail referred to its shape in the wind. In the seventeenth and eighteenth centuries different countries adopted different sizes and shapes for jib sails, by which they came to be identified. Thus, a Spanish ship would have a small jib or none at all, whereas a French ship would often set two jibs. Warships would have jibs cut thin so as to maintain point more easily—so a suspicious enemy sighting a slim foresail at distance might not like the cut of his jib.

Devil to pay: *The bad consequences of an act.* "Pay" means to seal the seams with tar or pitch. So, to sailors "paying the devil" meant caulking the longest seam with pitch. The devil was the most difficult seam to reach; in dry dock the sailor would often be required to squat in the bilges to do so. Hence an unpleasant task, perhaps allocated as punishment.

Loose cannon: *A reckless, unpredictable person.* Quite literally, a cannon that was loose, i.e., not properly secured. On the deck of a rolling man-o'-war such heavy ordnance could be very dangerous to crew and ship.

5

A Lot of Torque

Leeway: *Margin of freedom; permissible degree of error.* A lee shore is a shore that is downwind of a ship. Leeway is the space between ship and shore that a sailing ship must maintain to avoid being blown onto the shore.

Listless: *Weary; unwilling to make an effort.* A becalmed sailing ship would not heel over (list) or move.

Overhaul: *Repair; overtake.* In windy weather, sailors went aloft to slacken buntlines to prevent chaffing of sails—requiring lines to be hauled over the tops of the sails.

Overwhelm: *Bury; submerge utterly; overpower.* From Middle English *whelmen,* meaning "to turn upside down" and later "to submerge completely."

Torque about Helmsmanship

You have escaped work and domestic slavery for a weekend on the water. You have left your wife, if only for a couple of days, and are snuggling up to your sweetheart, *Puddleduck,* a modest, though spoiled and well-loved little sloop. You take her out on the water, give way to an irritatingly splendid looking schooner called *Sparrowhawk* (silly name!), triumphantly slip past an ancient, plodding, yet strangely appealing brig that labors under the ludicrous appellation *Snoozing Goose,* and then change heading, from close hauling to a beam reach. Contemplating life, the universe, and everything nautical, your mind turns lightly to thoughts of torque.

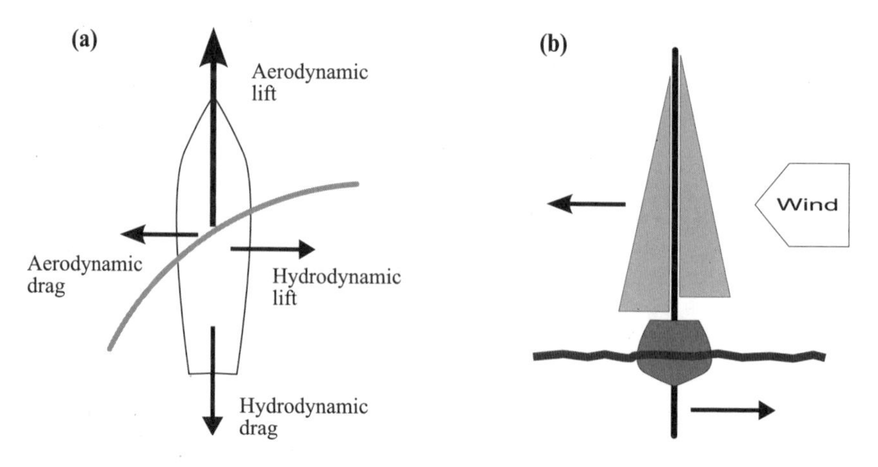

Figure 5.1. (a) *Puddleduck* on a beam reach. The wind generates aerodynamic lift and drag forces; hydrodynamic drag opposes boat movement through the water. Hydrodynamic lift arises from the keel, which acts as a hydrofoil. The lateral forces (aerodynamic drag and hydrodynamic lift) must be about the same to reduce leeway to a minimum. But these forces do not always apply at the same point, thus generating torque. (b) Seen from another angle, the two lateral forces also generate a torque (the heeling moment, here counterclockwise) about the longitudinal axis.

Huh? Well, you are a weekend sailor and these things are important. As you head across the wind you recall figure 5.1 of a coruscatingly brilliant book you once read about the physics of sailing.

We have already met aerodynamic lift and drag, two forces that are perpendicular to each other. Figure 5.1 shows a relatively simple case in which, because *Puddleduck* is on a beam reach, hydrodynamic lift opposes aerodynamic drag. If these two forces exactly cancel each other, *Puddleduck* makes no leeway. In practice some leeway is unavoidable, but careful design of the keel minimizes leeway. Hydrodynamic lift arises because the keel is a hydrofoil, an underwater wing, providing lift in the direction shown as it cuts through the water. Thus, your boat moves through the water because of aerodynamic lift, which then leads to hydrodynamic lift, which opposes aerodynamic drag and gives rise in turn to hydrodynamic drag. And this is a simple case.

You can also see in figure 5.1 that the two lateral forces, though they cancel each other in magnitude, generate a torque because they are applied at different points. Torque is a twisting force, and the torque experienced by *Puddleduck* in figure 5.1 has two components, which I will call the "heeling torque" and the "heading torque." Most clearly, the keel and the sails are widely separated, and so the lateral forces generated by these "wings" apply at different places: aerodynamic drag applies at the sails' center of effort (CE) (of which more shortly), whereas hydrodynamic lift applies at the keel's CE. The result, for the situation illustrated in figure 5.1, is a counterclockwise twist about the longitudinal axis of the boat, which you feel as the heeling moment. The second component of torque applies about a vertical axis parallel to the mast and changes a boat's heading because in general the hydrodynamic lift force applies slightly aft or slightly forward of the aerodynamic drag CE, as shown in figure 5.1. To see why this is so, and what the consequences are for your helmsmanship, we need to get to grips with CE and with the hydrodynamic drag equivalent CLR, or center of lateral resistance.

Center of Effort and Center of Lateral Resistance

The wind that impinges on a sail exerts a force at each point on the sail's surface. Lift is usually concentrated at the luff and varies with height. The total lift force on the sail is the sum of all these individual lift forces, taken at the sail center—hence the term "center of effort." We have seen already how to estimate the magnitude and direction of the lift and drag forces acting on a sail, but how do we know where they apply? The CE can be found quite straightforwardly for simple sail shapes such as triangles and squares. For triangles, there is a simple geometrical construction (shown by the dashed lines in fig. 5.2) familiar to any high school student.[1] For squares it is even easier: the CE is in the middle. For more complex sail shapes, and for combinations of different sails, calculating the location of the CE is difficult. There are methods of estimation,* but these are only approximate. In any case, the CE shifts every time you trim or reef a sail, and Mother Nature won't wait for you to get out a spreadsheet and work out the theoretical CE—she will blow the winds and fill your sails, and your boat will tell you more or less where the CE is located. Every helmsman knows that this matters, because the relative locations of CE and CLR have a big say in how your boat handles.

The center of lateral resistance, or CLR, is the geometrical center of the underwater part of the hull when viewed in profile. Again, this is

*Gerr (see bibliography) goes into this in some detail, for example, showing how to calculate the CE for a gaff-rigged yawl.

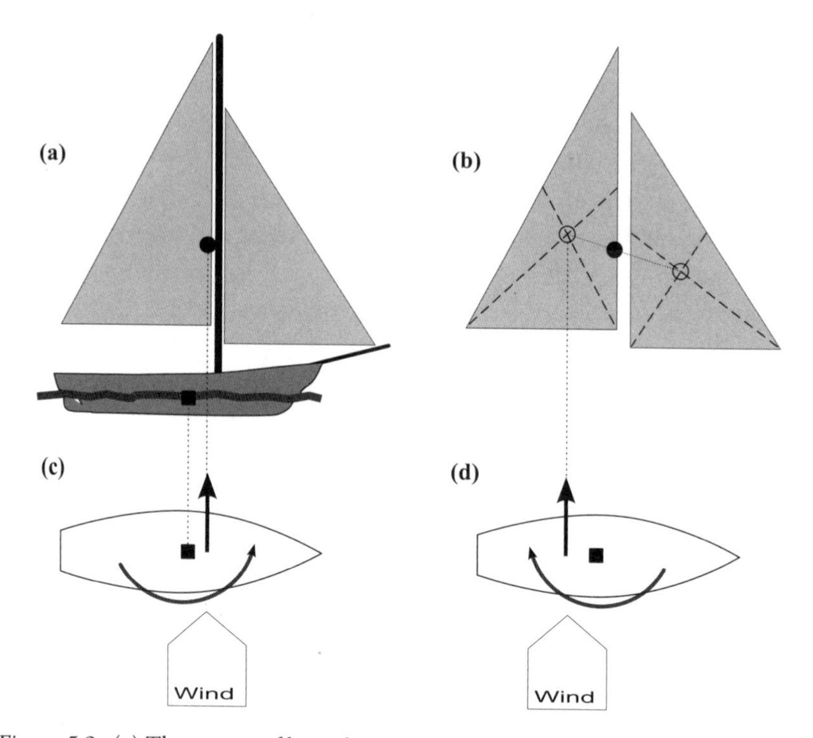

Figure 5.2. (a) The center of lateral resistance, or CLR (square), is slightly aft of the center of effort, or CE (circle), thus generating lee helm—bad news. (b) For simple sail shapes, such as triangles, the CE can be calculated without difficulty, but for more complicated sails CE is not easy to evaluate. (c) A lee helm; *Puddleduck* turns to follow the wind. (d) By furling her jib, the helmsman shifts *Puddleduck*'s CE aft of her CLR, and she is now on a weather helm—a much more desirable outcome.

difficult to calculate. For a small boat such as *Puddleduck*, however, CLR can be estimated quite simply. Stand on shore with *Puddleduck* parallel to the shore and push straight out, at a point near her middle. If her bow moves farther away from you than her stern, repeat the exercise with your push slightly closer to the stern. Keep doing this until she moves away from you without any rotation. Of course this rough-and-ready method will provide you with only an approximate CLR location, and anyway the CLR changes when your boat heels, because the shape and size of the wetted hull surface changes.

So now you know the whereabouts of CE and CLR. So what? The significance of these imaginary points for a helmsman is as follows. If the CE is aft of the CLR, *Puddleduck* is said to possess a *weather helm,* meaning that she will tend to turn upwind. With the wind on her beam and no action taken by you, torque will act to point her bow to windward. A small weather helm is generally reckoned to be a good thing but not when heading downwind because steering is difficult (she wants to turn around to face the weather). CE forward of CLR (a *lee helm*) is not good because it results in a lot of leeway. A *neutral helm* corresponds to CE and CLR in the same vertical plane, in which case no torque acts to turn *Puddleduck* toward or away from the wind. As helmsman you can influence CE directly by trimming sails, etc; CLR changes with heeling angle. Thus, the manner in which, without any action from you, your boat reacts to a gust of wind may change. For example, she may have a neutral helm when vertical and so heel over without changing course, but then, when heeling, the CLR may shift so that she turns to windward. Obviously these characteristics, and how they change, are important and always have been for all the different types of sailing vessels. Designers of sloops, ketches, yawls, or schooners will tell you where they would like the relative positions of CE and CLR for their vessels, and these positions will change between types of boats. Torque does not make the helmsman's task easier,* but it is a fact of life. Understanding the effects of torque may help a little; it may at least explain why your boat changes course the way she does, without your adjusting the rudder. A detailed theoretical analysis covering a wide variety of circumstances is next to impossible because of the number of complicating factors,† and so I will leave the subject of heading torque.

Before we turn to heeling torque, however, a final paragraph about CE

* Except it helps you to change heading while stationary (in which circumstance your rudder is useless). By furling or unfurling a sail you shift CE and so change the heading torque.

† Suppose the wind direction is oblique, rather than coming from the beam. How does CE change with heel angle or with sail trim? How does CLR change with a heavy load or in a heavy sea? Or with speed? Not only would calculations of these effects be difficult, but they would also be pointless because there is no precise way to act on the numbers produced. A helmsman develops an intuition about how his boat is going to handle in a given circumstance; this intuition is, I suppose, his subconscious analysis of all his experiences with the boat. He will know what to do when, say, he is on a broad reach at 10 knots in a choppy sea with only the jib flying. No theoretical calculation will be of any use to him at all. The physical principles expounded here (such as torque) explain the forces that arise but are not—and cannot possibly be—detailed enough to make precise predictions that are realistic in all circumstances.

and CLR that will emphasize the complexity of boat-handling dynamics. CE and CLR are two numbers that crudely represent the forces that act on the sails and the hull (including the keel). There is a complicated interplay between aerodynamic lift and drag above the waves, and hydrodynamic lift and drag beneath the waves, not to mention the influence of the waves themselves. I will later describe something of the hydrodynamics. Always bear in mind that the real-world operation of a boat involves an interaction among all of the component pieces in a scientifically choreographed manner of mind-boggling complexity. For example, you hoist a sail in a stationary boat; the wind causes the boat to move; the boat's movement, because of the hull shape and the keel, generates hydrodynamic lift and drag; lift and drag alter the boat's course; the change in heading, in turn, adjusts the angle of attack and so changes the aerodynamic lift; etc. At the same time the changing speed generates a bow wave which changes hydrodynamic drag and CLR, while wind and changing course direction influence heeling, thus altering aerodynamic lift . . . you get the picture.*

Heeling

The lateral forces of figure 5.1b clearly are going to apply a torque; the boat will heel over at an angle, which we will determine here. Small boats with high sails and little or no keel are particularly susceptible to heeling in a crosswind (fig. 5.3); as we will see, wind speed is the single most important factor that determines how much a boat heels. You may regard heeling as inconvenient, irritating, alarming, or dangerous, depending on the angle you find yourself heeling over. Clearly, as your mast tips over, the sails will spill air and generate less lift. Heeling may also cause *Puddleduck* to suddenly head up or make some other unwanted course change due to changing CLR and CE, as we have just seen. This is particularly problematic if you are racing or are restricted to a narrow stretch of water, but it is a part of sailing life. Indeed, to reduce the amount of leeway made by your boat, she may have a keel or center-

*Carl Chase has a very good nonmathematical description of the interaction of sail and keel forces. The scientific complexity of these interactions is not, as with fluid dynamics (aero- or hydro-), because the fundamental physics is difficult. On the contrary, the ideas underpinning torque are basically quite simple and intuitive. The complexity arises because of the number of interacting torques, acting in different directions.

Figure 5.3. Sloops heeling in a crosswind. Image from Wikipedia.

board, which, by the very act of preventing leeway, induces heeling. Heeling is the price that a sailboat pays for reduced leeway.

There are a number of actions that *Puddleduck*'s intrepid helmsman (that's you) may take to reduce heeling. You might trade in a bit of leeway by raising her centerboard; the increased leeway will reduce apparent wind speed and thus reduce heeling angle. If *Puddleduck* is a larger yacht with a canted keel, this design feature can be applied to reduce heeling angle, as we will see. You and your crew might trapeze—hang off the side of the boat, thereby adding your torque to that of the keel—on *Puddleduck*'s weather side to directly counter the heeling torque of the wind. You can sheet out the sails to spill wind, or flatten them to reduce lift, depending on circumstances. Lastly, of course, you might simply point her bow to the wind, trading forward speed for a more comfortable upright position.

We can understand quantitatively how much heeling occurs with a simple approximate physical model. Figure 5.4a depicts a heeling boat with a keel for which I have assumed that the wind force is applied at the CE, which is a height, h, above the longitudinal axis of rotation (the axis around which the boat heels). Similarly, the keel weight is effectively at a depth, h_{keel}, below this same axis. To simplify the calculations that follow, I assume that the boat hull has a cross section that is more or less circular, as seen in figure 5.4.[2] The heeling angle is the angle, a_{mz}, between the mast, m, and the vertical direction, z, of figure 5.4. A boat heels over until an equilibrium heeling angle is reached; this angle is then constant until something changes, such as wind speed or heading. By applying Newton's laws, we find the equilibrium heeling angle to be[3]

$$\tan(a_{mz}) = \frac{\rho A h w'^2}{m_{keel} g h_{keel}} \tag{5.1}$$

This simple analysis explains much about heeling. As we would expect, the heeling angle increases as effective sail area, A, increases, and as CE height, h, increases. It increases as the square of apparent wind speed, making wind speed the single most influential factor in the magnitude of the heeling angle. A heavier, longer keel reduces the heeling angle. From equation (5.1) we can see that, if a boat heels 5° in a 5-knot wind, then it will heel at 19° in a 10-knot wind (and 38° and 54° in 15- and 20-knot winds). Equation (5.1) also explains why bringing up the centerboard might reduce heeling. Such an action would reduce keel depth, h_{keel}, and apparent wind speed, w'. So long as the combination w'^2/h_{keel} is smaller than it was, the heeling angle is reduced. A trapezing crew will have the same effect as increasing the keel mass, m_{keel}.

Finally, it is worth noting how the helm of a sailing vessel can be changed as a result of heeling torque. Figure 5.4b shows a boat from above. When upright she carries a lee helm, but when she is heeling to starboard, the torque changes to provide her with a weather helm. This is because heeling moves the CE beyond the deck and out to sea, as viewed from above. Here, we have two torques combining: the heeling torque twists the boat hull about her longitudinal axis, whereas the CE-CLR torque twists her about a vertical axis, either clockwise or counterclockwise, as shown.

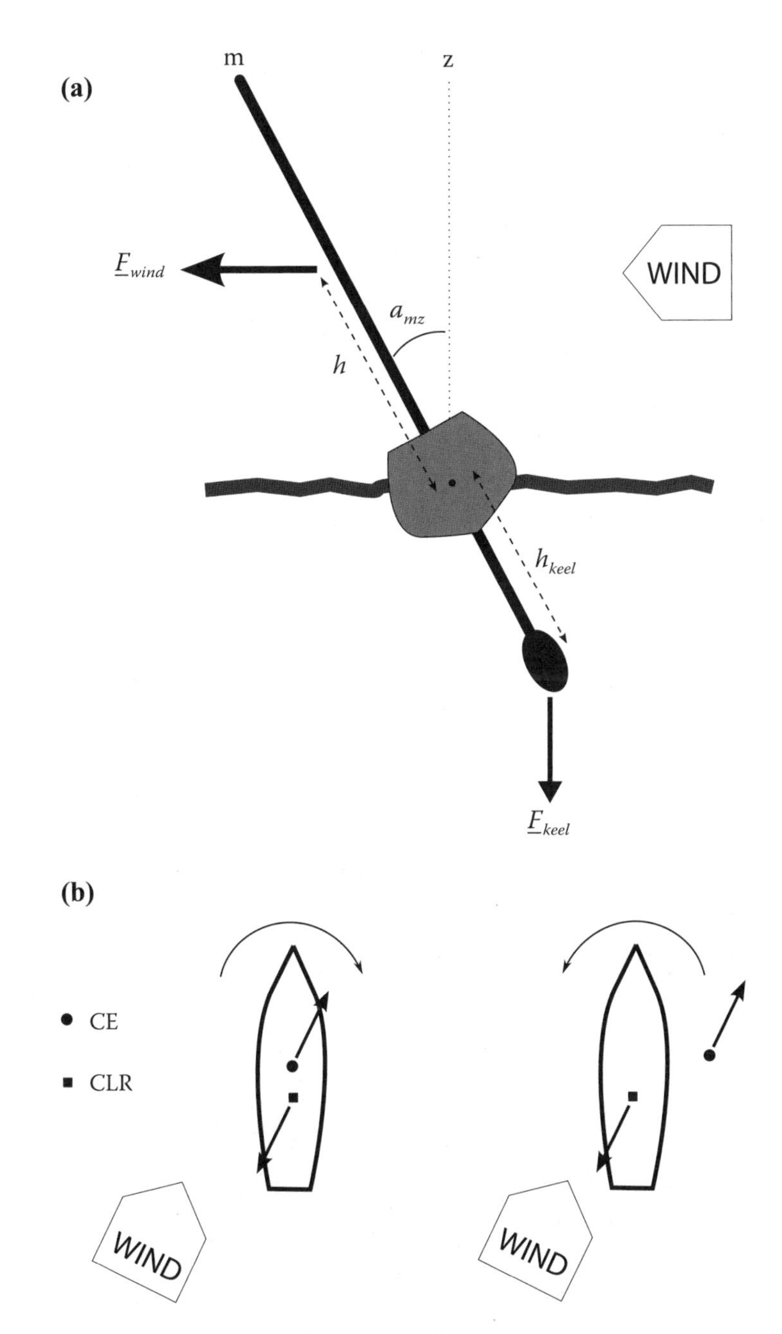

Figure 5.4. (a) Wind force applies a counterclockwise heeling torque, whereas the weight of the keel exerts a clockwise torque. These opposing torques balance at the equilibrium heeling angle, a_{mz}, calculated in the text. (b) Heeling can change a boat's helm.

Torque about Hull Stability

Torque applied about the longitudinal axis leads to heeling, and if the heeling angle is large enough, a boat may capsize. This is considered to be undesirable among all the boat owners I know, and so hull stability is taken to be a matter of some importance. Here, stability refers to the ability of a boat to right itself after having been heeled over by, say, a sudden gust of wind or a sudden change of course. We would like to know how much our boat can heel over before capsizing, and we would like to know how fast she can right herself. I will deal with these questions here. The physics of hull stability is simple in concept but devilish fiddly when it comes to detailed calculations. So, in the spirit advocated in the introduction, I will once again simplify when necessary to make the issues clearer and will relegate math details to the endnotes. The fiddly details have generated much literature, and a number of technical terms have entered the sailor's vocabulary as a consequence. So that you know what is being talked about when these words are uttered, I will begin by introducing them.

Center of Buoyancy and Metacenter

The force of gravity that acts on an extended object, such as your *Puddleduck* or indeed yourself, can be thought of as acting at a particular point, called your *center of gravity* (CG). This is a familiar idea, and it extends to other forces. Thus, the force that buoys up your boat acts on all parts of the wetted surface but can be thought of as acting at a single point called the *center of buoyancy* (CB). So far, so good, but as soon as we think about CB a little more deeply, we see that it quickly becomes complicated. Consider figure 5.5. Here, for simplicity I have replaced your beloved *Puddleduck* with a block of wood—no insult intended—but you will appreciate the reason for this substitution shortly. The block of wood is of such a density that the water line is halfway up when the block is floating in a stable position. The CG is easy to locate: it is simply the geometrical center of the block, assuming uniform density. What about the CB?

The center of buoyancy for a simple shape such as this block is the geometrical center of that portion of the block which is underwater. So, for the initial stable position we can readily mark the CB with an x, as in figure 5.5. Now let us suppose that the block is disturbed and rotates a

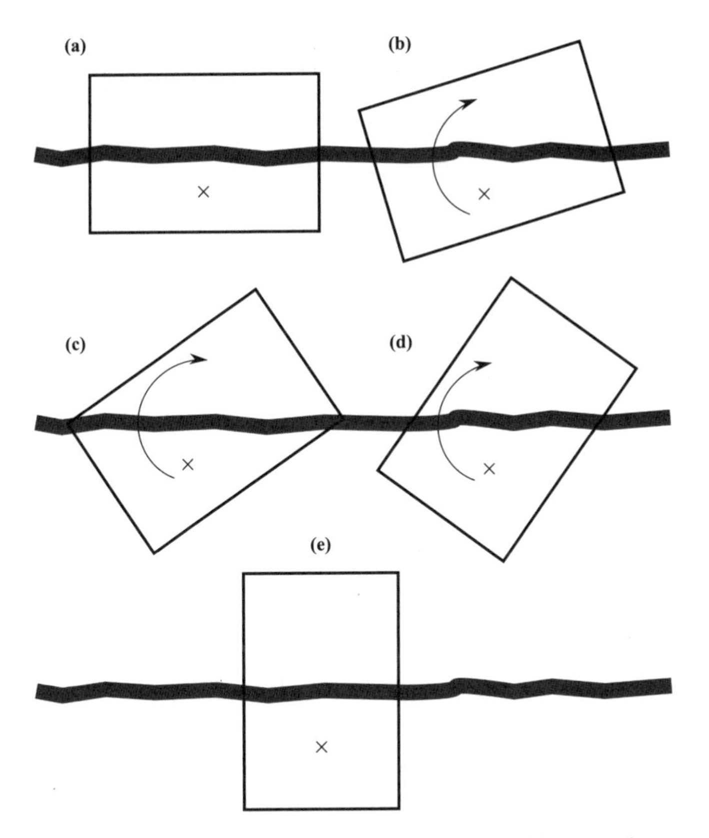

Figure 5.5. (a) A floating block in a position of stable equilibrium. The center of gravity, CG, lies in the same vertical plane as center of buoyancy, CB (x). (b) When the box is rotated, its CB moves and is no longer in the same plane as the center of gravity, resulting in a restoring torque. (c) and (d) For increased rotation angles, the CB changes, thus altering the torque that is applied to the box. (e) When the rotation angle is 90° the CG and CB are again aligned and no torque applies. But this position is one of unstable equilibrium.

little, like a heeling hull. This new orientation does not alter the CG at all, but the CB has moved. We can calculate the new CB easily enough for a simple block of wood, but for a complex hull shape it can be impossible to work out with a pen and paper: we would need to resort to number-crunching, which would muddy the pedagogical waters. I show the CB for our block of wood for several different orientations in figure 5.5. That the CB depends on heel angle, whereas the CG does not, has the following consequence. When the two centers are not aligned vertically, the force of gravity and the force of buoyancy (which have the same magni-

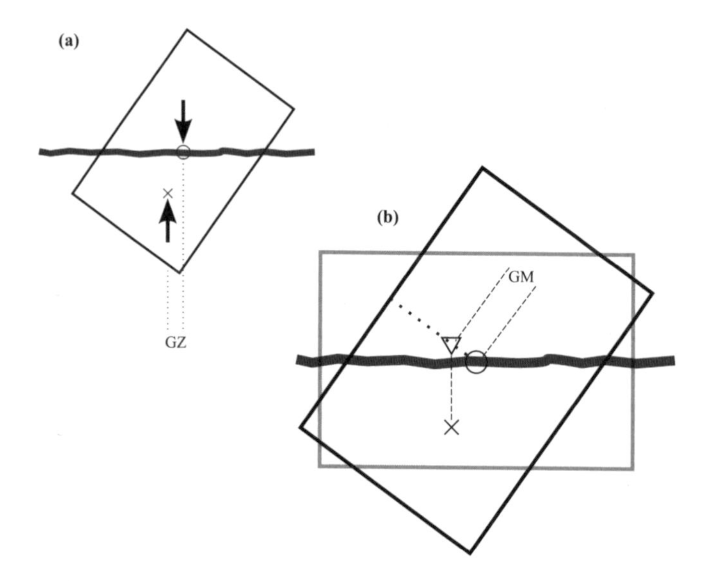

Figure 5.6. (a) Center of gravity, CG (open circle), and center of buoyancy, CB (x), for a floating box. If the box is a solid block of wood, the CG is at the geometrical center, in this case chosen to be at the water line. Arrows indicate upward buoyancy force and downward gravity force. These forces are not aligned and therefore apply a torque to the box. The horizontal separation of CG and CB is the righting arm, denoted GZ. (b) If the box was originally in a stable position (outlined in gray), a line from the CG to the top would appear on the rotated box as shown by the slanted dotted line. This line intersects the "new" vertical from the CB at the metacenter (triangle). The distance between CG and the metacenter is metacentric height, denoted GM. If the metacenter is higher than the CG, as here, the box orientation is stable—it will return to its original position. (The righting arm, GZ, is defined as positive or negative depending on whether the metacenter is above or below the CG.)

tude but opposite direction) do not quite cancel out but instead exert a torque on the hull. This torque also depends on the heel angle, which complicates the physical analysis significantly.

We can quantify this dependence of the buoyancy force on the heeling angle in a manner that readily conveys information to the eye, and so is very popular among boat designers. Much of a boat's heeling characteristics can be captured in a graph called a *righting arm curve*. The righting arm is the horizontal distance between CG and CB, as shown in figure 5.6. It is conventional to denote righting arm by the abbreviation *GZ*. This distance is important because it is proportional to torque:

increased GZ means increased torque. If GZ is positive then the torque is a righting moment that acts to reduce a boat's heeling angle; if negative then the torque will act to capsize the boat.

How do we decide whether GZ is positive or negative? In other words, how do we know if the torque that is applied by the combined forces of buoyancy and gravity will act to right a boat or capsize it? The answer lies in yet another technical term: *metacenter*. The metacenter is explained in figure 5.6. The concept is slightly abstract, so I will provide a separate definition of it here. Consider *Puddleduck* sitting calmly on a flat sea, with her mast vertical. A large wave comes along that causes her to heel over a large amount, as happened to the block shown in figure 5.6. *Puddleduck*'s center of buoyancy moves because when she is heeling, her hull displaces water in a different way than when she is upright, as we have seen. We find her metacenter at the new heeling angle by a simple geometrical construction. The CB acts vertically: we extend a vertical line from the CB to a line through her mast. (Of course, now that she is heeling, her mast is no longer vertical.) The point of intersection of these two lines is the metacenter, denoted by a triangle in figure 5.6.

Righting Arm Curves and Stability

Here is the significance of the metacenter: if it lies above the center of gravity, the torque will right *Puddleduck:* she will return to her stable upright position. If the metacenter lies below the CG, the torque will cause her to capsize. Of course, the location of the metacenter changes with the heeling angle because the CB changes; that makes it difficult to say what is going to happen by quickly glancing at the equations. However, the stability information is readily conveyed in *Puddleduck*'s righting arm curve, which is a plot of GZ vs. heeling angle. Determining the righting arm curve for a complicated structure like a boat is difficult and depends on every last detail of hull shape and mass distribution. So instead, in figure 5.7, I show the righting arm curve for our block of wood. You can see that the shape of the curve changes as the shape of the block changes. This change is observed also in boats: wider boats show more initial stability, meaning that they are more difficult to tip over. The righting arm curve displays this initial stability in the steepness of the slope near zero heeling angle, $a_{mz} = 0°$. A more stable block or boat has a steeper slope because it has a greater righting moment and so is harder to tip over. Catamarans are very wide and have great initial

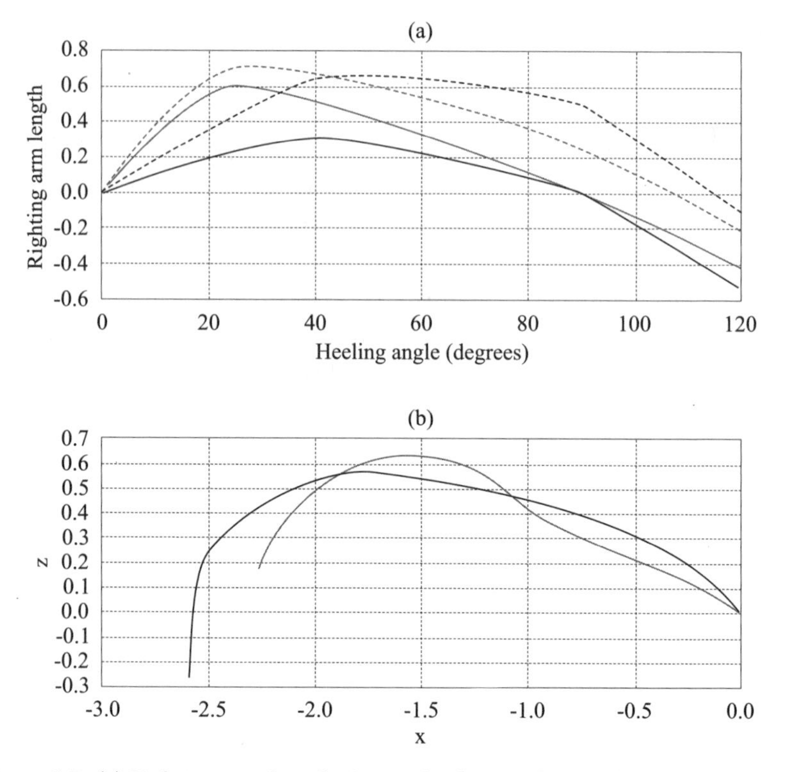

Figure 5.7. (a) Righting arm length, GZ, vs. heeling angle, a_{mz}, for different floating boxes. If the center of gravity of the box is lowered from the geometrical center (appropriate for a solid block) halfway to the bottom (more appropriate for a boat), the dashed curves result. Black lines correspond to box width and height of (w,d) = (3,2), and gray lines to (w,d) = (3,1). (b) Movement of CB from (x,z) = (0,0) as the heeling angle increases from 0° to 120°.

stability. More slender boats with deeper hulls, like our block of wood in figure 5.6 (with its righting arm curve also plotted in figure 5.7), have less initial stability. They tip over more easily and, as we will soon see, roll more; their roll amplitude and roll period (the time taken to roll back and forth once) are both greater than for a flatter boat, or block.

Note from figure 5.7 that the righting arm curve peaks at a certain heeling angle, corresponding to maximum righting moment, and then decreases as the heeling angle increases further. At these larger angles the boat or block will still try to right itself, so long as the righting arm curve is positive, but the torque is reduced. At some larger heeling angle the curve reduces to zero. At this point there is no righting moment, and the boat or block will stay at this angle if placed there: there is no torque to cause it to turn either way. At still larger heeling angles the curve is negative, corresponding to an overturning or capsizing torque; when a boat or block gets into this region, it cannot right itself. The heel angle at which the righting arm curve dips to zero marks the range of stability for our boat or block. The two characteristics that emerge from righting arm curves—initial stability and range of stability—are key design parameters for boat-builders. Obviously, we would like both to be as large as possible, but it doesn't work like that. I have tried to indicate in figure 5.7 how boat curves differ from those of wooden blocks by lowering the CG of the blocks. This is what would occur if the blocks were hollowed out to form a crude boat hull. The results show as better righting arm curves: greater initial stability, increased righting arms and hence increased righting torque, and greater range of stability. From figure 5.7 we see that flatter, hollow blocks have greater initial stability but a lesser range of stability, and the same is generally true of boats. The message is: go to sea in a boat, not a block—but I guess you already knew that.

Boats differ from blocks in other ways too. The downflood angle is the maximum heeling angle that a boat can have before she swamps. For an open boat, this angle can be quite small: the righting arm curve abruptly ceases instead of smoothly varying out to large heeling angles, as in figure 5.7. The watertight deck on many modern yachts permits much larger heeling angles. Even here, of course, it is necessary to close the hatches. Hatches tend to be placed along the centerline of a yacht, and as high up as practicable, to permit large heeling angles without danger of swamping. Another difference between boats and blocks is much more difficult to quantify. Boats carry movable masses which can shift as the boat heels. Whether it be gas or water in tanks, or cars and trucks on ferries, shifting mass can be very dangerous in a rolling boat. What seems stable can become unstable very quickly, and in the past a number of tragedies have resulted for this reason. To avoid the stability problems that can result from liquid sloshing back and forth, fuel tanks on yachts have baffles placed in them.

Time to Rock and Roll

Designing boat hull shapes for stability involves a number of trade-offs. Not least is the comfort of the crew and passengers. Not only must our

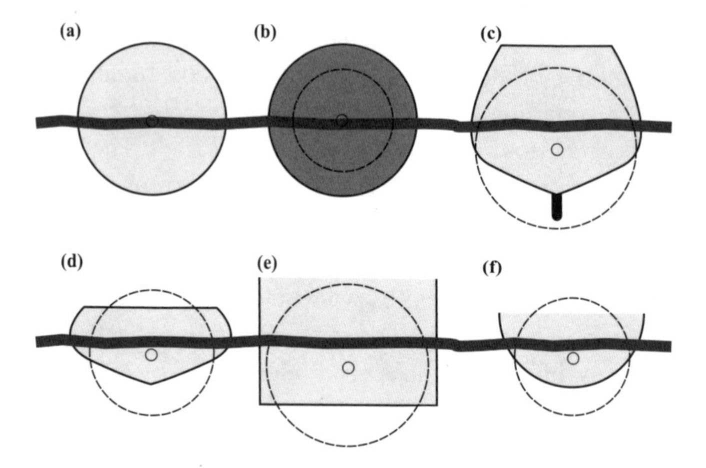

Figure 5.8. The radius of gyration (dashed circles) about the center of gravity, CG (small circles), of a boat hull for different hull shapes. If the hull is a hollow cylinder (a), the radius of gyration is just the cylinder radius. For a solid cylinder (b) the radius of gyration is only 71% of the cylinder radius. For different boat hull cross sections (c)–(f) the radius of gyration is different, as shown.

boat be stable, but she must be compatible with life aboard. A flat boat with a stiff response to heeling—one that rights herself sharply—may buffet those aboard, perhaps dangerously. Passengers may be knocked about as their yacht rolls back quickly to her upright position. On the other hand a tender response—from a boat that rights herself very slowly —may induce unease. If you are on board such a vessel, rolling slowly from side to side over large heeling angles, you may wonder whether or not she is ever going to come back or if she has passed the point of no return. Or you may simply be wondering whether or not you can hold down your breakfast. So the rolling behavior of a boat is important. Rolling is a consequence of stability, of course. We have seen how *Puddleduck* rights herself in response to heeling over. She has inertia, and the righting does not cease when she gets herself vertical. She overshoots and rolls the other way before coming back again. The resultant rolls will diminish in amplitude as hydrodynamic frictions damps out the motion (pardon the pun).* In this section I will investigate the rolling motion of a boat.

*Amplitude here means maximum heeling angle. So the maximum heeling angle is reduced with each roll due to friction.

First, I need to introduce yet another technical term.* *Radius of gyration* may sound to you like the size of a dance floor, but to an engineer it has a more prosaic meaning. Figure 5.8 depicts the profiles of a number of boat hulls and other floating objects. These objects may be induced to roll about their CG, and we need to know the effective radius of the object, as measured from the CG. For a hollow cylinder centered on the CG this is a no-brainer: the radius of gyration, R_g, is just the radius of the cylinder. For a solid cylinder of the same mass, however, it turns out that R_g is only 71% of the actual radius. This is because the cylinder mass is distributed over many radii, in this case from the CG (radius zero) out to the actual, physical cylinder radius, R. The "average" radius, when it comes to rotations, is $R_g = 0.71R$.[4] For more complex shapes such as a boat hull, of which several are shown in figure 5.8, the radius of gyration is difficult to work out algebraically; we must resort to a computer. I don't intend to do that here; so long as you appreciate that there *is* a radius of gyration for any rotating object, then I have got the message across.

Roll Time

Now we will calculate the period of a roll (a.k.a. roll time) for the boat profile shown in figure 5.8f. This profile is a half-cylinder of radius R, and we may assume that it is a reasonable approximation to the shape and mass distribution of many open boats. Because we are dealing with simple, approximate solutions, rather than complex, exact simulation results, this half-cylinder shape will serve admirably for the purpose of calculating roll time. It has a couple of features that reduce the math to a manageable form. In figure 5.9 we have the half-cylinder boat rolling on a calm sea. The roll angle induces a torque which tends to right the boat. If she is heeling counterclockwise, as shown, the torque acts clockwise, and vice versa.

We invite Sir Isaac Newton to cast his expert eye over the scene, and he obliges us by quickly deriving the equation of motion, which tells us how the boat rolls. We can easily solve the equation, and determine the roll time,[5] which is

*If you supposed that the plethora of technical terms to do with boat torque reflects the considerable amount of theoretical effort that has been put into understanding this subject, you would suppose right.

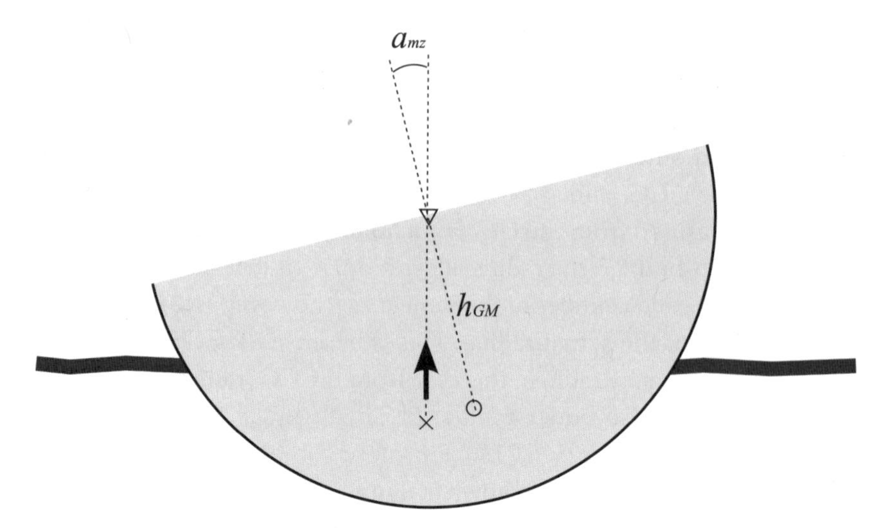

Figure 5.9. For a half-cylinder hull shape on a flat sea, the center of buoyancy, CB (x), does not change with roll angle, a_{mz}, and the metacenter (triangle) is always at the cylinder axis. The buoyancy force (arrow) applies a torque about the center of gravity, CG (open circle), and from this force we can derive and solve the roll motion equation. The metacentric height is noted by h_{GM}.

$$T = \frac{2\pi R_g}{\sqrt{gh_{GM}}} . \tag{5.2}$$

Equation (5.2) tells us that the time it takes our half-cylinder boat to make one complete roll does not depend on the size of the roll: it can be 5° or 25°, but the roll time is the same. Roll time does increases with the size of our boat, which is no surprise. One rule of thumb used by boat designers is that the roll time for any boat should be pretty close to the beam, meaning that the roll time in seconds should be about equal to the beam dimension measured in meters. The reason for this rule is crew and passenger comfort. Faster roll times create jerky, harsh motion, while slower roll times can induce queasiness. For a comfortable boat ride the hull should be built with a metacentric height, h_{GM}, and radius of gyration, R_g, to comply. For our half-cylinder we have $R_g \approx h_{GM} \approx 0.8R$, where R is the physical radius of the boat. Plugging this radius of gyration into equation (5.2) gives us a roll time of 1.6 seconds for a boat radius of 0.8m (and so a beam of $2R$ = 1.6m). So, the requirement for a comfortable outing in our half-cylinder boat constrains the boat beam to be between 5 and 6 ft.

Real boats are more complicated to analyze, but the same kind of results apply. Designers must balance a number of different requirements to produce a good boat. The metacentric height and the radius of gyration are juggled with the desired boat beam to yield an acceptable roll time.

Modern boat designs can "cheat" equation (5.2) by introducing other influences. For example, a large keel will dampen down rolling motion, as we will see, so that any discomfort that results will be of brief duration. Some large and expensive boats have a keel that can be canted. It can be set out to one side or another to counter rolling by increasing the amount of righting torque. This is an expensive and not altogether satisfactory solution,* but it shows the lengths to which designers are prepared to go in order to reduce roll. Another solution that has been considered is to install movable ballast. Imagine a heavy weight that can be shifted dynamically from one side of your boat to the other, with the movement timed to provide maximum righting torque. Such a dynamic solution would require computer-controlled equipment for shifting the weight in a timely manner. An altogether simpler solution, in wide use, is to deploy flopper-stoppers—floats that dampen rolling motion. These work only when the boat is moored, but they cut out the big rolls that can result from, say, the wake of a large passing vessel striking you on the beam. Flopper-stoppers work by significantly increasing the damping effect of friction (fig. 5.10).

Resonance

Puddleduck is in harbor one calm evening, and you are entertaining a lady friend on board. A large cruiser sails silently by. You pour a glass of red wine. The cruiser's wake hits *Puddleduck* on the beam, and you spill the wine over the lady friend. She is about to remonstrate about her ruined white dress when the next wave hits, tipping her onto the floor

*It is technically difficult to install an effective keel that moves in this way. Also, a canted keel is less effective at resisting leeway motion. So here we have another example of "trade-off": different design requirements tug boat designers in different directions. Good design is all about successful compromises.

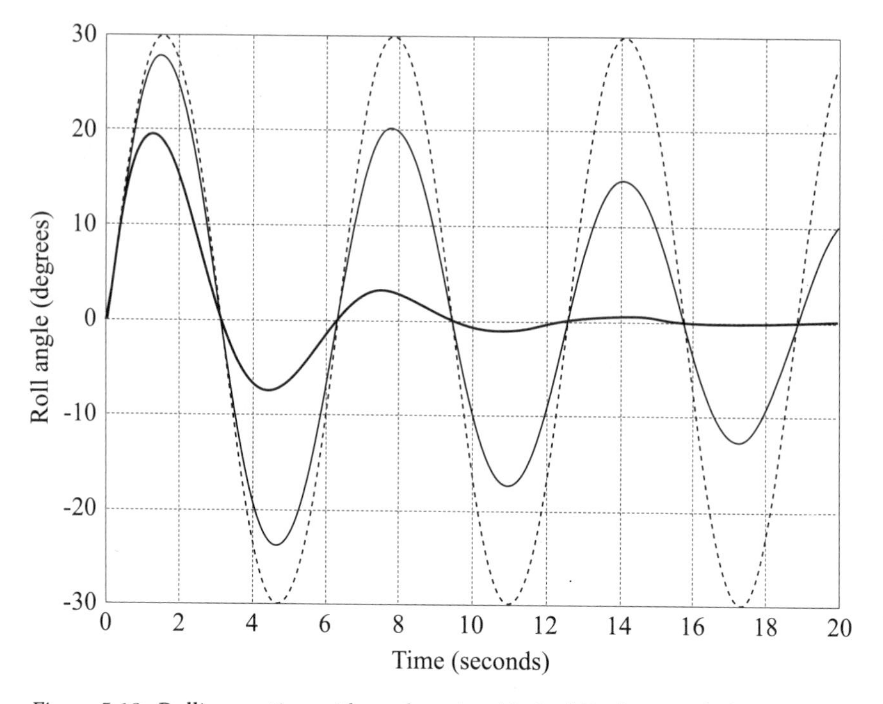

Figure 5.10. Rolling motion with no damping (dashed line), more realistically with a little damping (thin line), and with flopper-stoppers deployed (bold line). These curves have been simulated—they are not experimental results obtained by measuring rolling boats, and the time axis is arbitrary—but they capture the physics of what is happening. The flopper-stoppers reduce the duration, as well as the peak amplitude, of the roll.

and you on top of her. She is about to remonstrate about inappropriate behavior when the next wave rattles you both around *Puddleduck* like two peas in a can. She is about to leave when the next wave tips her overboard. OK, so my humorous introduction to rocking and rolling is also a little overboard, but I hope that you can appreciate the wave phenomenon epitomized here.

Your thoughts, on seeing your lady friend tipped overboard by a series of waves, as she shrieks for help and splashes frantically in the water, naturally turn to wave resonance. A series of waves, particularly if they approach a boat broadside on, can instigate rolling behavior that is much more severe than that induced by a single wave. The phenomenon of resonance is at work here. One boat in the harbor may rock violently back and forth, while none of the others seem much affected. It is like the wineglass (empty, this time) that is shattered by a diva's voice, but only if the pitch is just so.*

I can demonstrate this familiar wave resonance phenomenon to you by again making use of the half-cylinder boat. This time, as you can see in figure 5.11, I have given her a keel. The reason for doing so is to emphasize the role played by friction in the resonance phenomenon. As with flopper-stoppers, a keel helps to mitigate the effects of rolling motion by increasing the effective friction between the boat and the water. It is straightforward to set up a simple (approximate) mathematical model and solve the equation of motion for such a boat influenced by a series of waves. First, we model the waves by a sinusoid of specified wavelength and speed, and then let loose Sir Isaac once more to tell us how the boat's rolling angle changes when she is hit broadside by these waves.[6] The answer is that the rolling angle changes in time as follows:

$$a_{mz}(t) = a_0 \cos(\omega t - \phi) + transient \tag{5.3}$$

A transient is a disturbance that settles down quickly, leaving a more persistent solution. I will ignore the transient behavior here. What is left, once the transient effects have dissipated, is a rocking motion with angular frequency, ω and amplitude, a_0.† The quantity φ is a phase factor, which describes by how much the rocking angle of the boat lags behind the water wave. This phase angle is predicted by my model but is not of much interest here, so let's ignore it. The angular frequency and amplitude can be expressed in terms of water wavelength and boat parameters:

$$\omega = \sqrt{\frac{2\pi g}{\lambda}}, \; a_0 = \frac{\Omega_l^2}{\sqrt{(\Omega_0^2 - \omega^2)^2 + b^2\omega^2}} \, . \tag{5.4}$$

Here λ is water wavelength, b is friction coefficient as before, and

$$\Omega_0 = \sqrt{\frac{gh_{CG}}{R_g^2}}, \; \Omega_l = \sqrt{\frac{2\pi h}{\lambda}\frac{gh_{CB}}{R_g^2}} \, . \tag{5.5}$$

We have already met radius of gyration, R_g. The CG and CB distances h_{CG}, h_{CB} are constant for our half-cylinder boat and are defined in figure 5.11. Water wave height is h.

*At this point you dive in, rescue lady friend, send her tearfully home in a cab, and return to thinking about wave resonance.

†Angular frequency is frequency divided by 2π, so an angular frequency of 6.28 sec^{-1} is the same as 1 Hz, or 1 cycle per second.

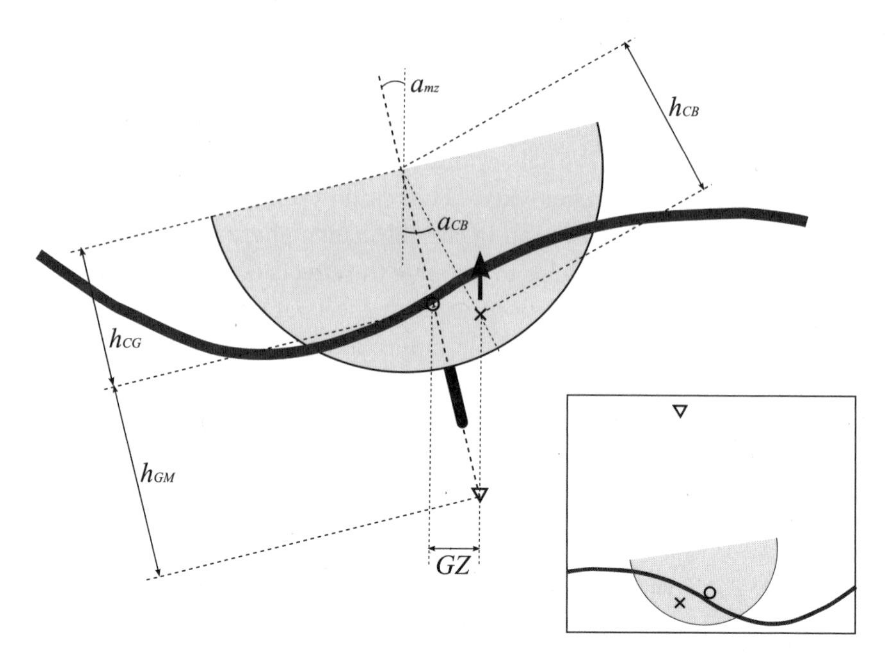

Figure 5.11. For a half-cylinder hull hit by a broadside wave, the buoyancy torque is as shown by the arrow. The righting arm, GZ, is the horizontal distance between the center of gravity, CG (open circle), and the center of buoyancy, CB (x). Metacentric height is h_{GM}. Roll angle is a_{mz}, and wave slope angle is a_{CB}. Note that the metacenter lies below the CG, and so the torque in this instance tends to capsize the boat. Note also the distinction between CG distance, h_{CG}, and metacentric height, h_{GM}; they were the same in fig. 5.9, but not here. *Inset:* A short time later the metacenter is above the CG, and so the torque acts to right the boat. The metacenter moves wildly, resulting in uncomfortable rolling motion.

So what do these equations tell us about boat rocking due to a series of waves on the beam? Any physicist looking at the form of equation (5.4) for amplitude a_0 would immediately cry out "resonance!" The amplitude can vary over an enormous range depending on boat and wave characteristics. In detail:

- As you would expect, the boat rocks at the wave frequency.
- The rocking amplitude is at a maximum when $\Omega_0 = \omega$, which is to say when the water wavelength is $\lambda = 2\pi R_g^2/h_{CG}$. When the wavelength equals this combination of boat hull gyration radius and CG height, the boat resonates with the wave, producing maximum rocking action. If the boat hull parameters are different, rocking amplitude is

less. This difference in hulls explains why some boats are affected more than others by the same set of waves.

Given that resonance occurs, the amplitude of the boat roll is found to equal the following messy expression: $a_{0,max} = \sqrt{gh_{CG}}h_{CB}h/bR_g^3$. Let me try to simplify this a little. We expect that h_{CG} and h_{CB} are both proportional to R_g, and so we can say that the maximum roll amplitude behaves like $a_{0,max} \sim h/bR_g^{3/2}$. Bigger waves mean bigger rolls, unsurprisingly. Increasing friction helps reduce roll amplitude, so deploy those flopper-stoppers. Most significantly, increasing the radius of gyration significantly reduces roll amplitude. Thus, bigger boats suffer much less than smaller ones.

Everything I have said in this section applies to monohulls. We can perform a similar calculation for catamarans, and as you would expect, the results are rather different. This section is already technical enough, so I won't extend it with yet more math but will simply summarize the outcome. We obtain resonance behavior for catamarans in the same way as for monohulls, but the peak amplitude is smaller in general because catamarans are wider. Indeed, if the wavelength happens to equal the catamaran hull separation, there is no rolling motion at all; the cat just rides up and down with the waves. The conditions for resonance, when it does occur, are very different from the monohull case. Here, the key factor is the average hull depth, d_0, under water. Resonance happens when the water wavelength is $\lambda \approx 2\pi d_0$. So we see that a flotilla of cats at anchor will respond differently to the wake of a passing cruiser. Some will agitate violently, and other will not.

Stop Torquing

I have touched on several aspects of torque as it applies to sailing boats. We have seen how torque can change heading direction, unintentionally (for example, while the boat is heeling) and intentionally (when the helmsman adjusts sail trim). Another torque, perpendicular to the heading torque, applies to heel the boat and to cause it to roll; we have seen how the basic physics works here. As I have emphasized repeatedly, the details are much more complicated than I have portrayed with my broad-brush approach, but these details hide the basic principles.

My technical discussion so far has included two chapters on the subject of sails and one on torque. The next chapter dives underwater to deal with the physics of sailing that you cannot see: the action of keel and rudder and how your boat flies underwater.

Barge in: *Intrude suddenly.* Flat-bottomed barges were notoriously difficult to control and frequently bumped into other vessels.

By and large: *Overall, on the whole.* Sailing into the wind is referred to as sailing "by the wind," whereas a "large wind" is one almost from astern. So a ship that sails "by and large" performs well in all directions.

Hand over fist: *Continuous rapid advancement.* Originally "hand over hand," expressing literally the manner in which British sailors in the Age of Sail ascended the shrouds. The change to "hand over fist" is thought to have been made by U.S. sailors.

On your beam ends: *Hard up, in a bad situation.* A ship was on her beam ends when lying over so much that the deck beams were almost vertical.

6

Flying through Water

Flotsam: *Debris floating on the surface of the sea.* Material lost accidentally overboard—for example, deck cargo lost in heavy seas.

Jetsam: *Debris thrown overboard.* Material intentionally jettisoned overboard—for example, to save a ship from foundering.

Junk: *Discarded but possibly useful material.* From Middle English *jonk.* On a sailing ship, an old line (cable or rope) that could no longer support a load, but was retained for other purposes such as making a fender.

Pooped: *Exhausted.* A ship is pooped when her high stern section (poop) is swamped by a following wave.

Just about now you may feel a little righteous indignation toward your well-meaning author. He started me off well, you may say, in a Greek trireme, before graduating me to an imaginary sail aboard a caravel bound for terra incognita across distant oceans, culminating in a terrifying yet totally exhilarating race around the Cape, perched precariously on a skysail yard of a clipper bound for London from the exotic Orient. But it has been all downhill since chapter 3. In the last chapter my modest sloop *Puddleduck* was turned into a block of wood. What next?

Well, all I can say is no pain, no gain. It will get worse before it gets better. In this chapter I will start you off in a raft, and a particularly stupid kind of raft at that. Only an academic landlubber could think of designing such a graceless vessel as you are about to board because it will prove to be far from seaworthy. Indeed, the only merit of this craft is in demonstrating the phenomenon of hull speed, which is my first topic.

This chapter deals with the sailing physics that occurs on and under the waves. I will start and finish on the surface, with the effects of water waves on boat hulls and vice versa. In between, we will venture below the surface to examine the considerable influence of the keel and the rudder on a boat's sailing performance.

Hull Speed

Hull speed is a phenomenon of displacement boats, and not of planing boats. Most sailing boats and all ships displace water—move it aside—as they plow through it. Planing craft, such as most motor boats, glide over the top like a surfboard. It takes more energy to push water aside than it does to slide over the top of it, and so displacement boats move at a more sedate pace than their lighter planing cousins. Some small sailing boats can be made to plane, but the general rule is that sailing boats are of the displacement type. Hull speed is usually an upper limit to the speed of displacement boats.* It is unsurprising that such a limit exists: we have seen how drag increases with speed, and so sooner or later drag will balance out the drive force and a sailboat will not be able to go faster. Yet there is a surprise in store for those of you who are not familiar with sailing: the hull speed of a given boat depends on its hull length at the waterline. It is not obvious from a simple consideration of drag why this should be so, but it is a well-attested fact, often quoted in the sailing literature, that the maximum natural speed of a displacement boat (in knots) is $^4/_3$ the square root of waterline length in feet.

A key feature of the phenomenon, again well known to any sailor, is that hull speed has been reached when the bow wave of the boat lengthens to the waterline length. At lower speeds, there may be three or four complete waves seen to lap along the boat hull, but this number decreases as the boat picks up speed and reaches, pretty closely, one complete wave by the time the boat reaches her hull speed. It may be possible for her to go faster than hull speed, but this requires a disproportionate amount of effort. In other words, the hydrodynamic drag

*There is one trick by which a small displacement boat can exceed hull speed without expending enormous effort, and that is by surfing. Riding along the front of a wave is not the sole preserve of surfboards.

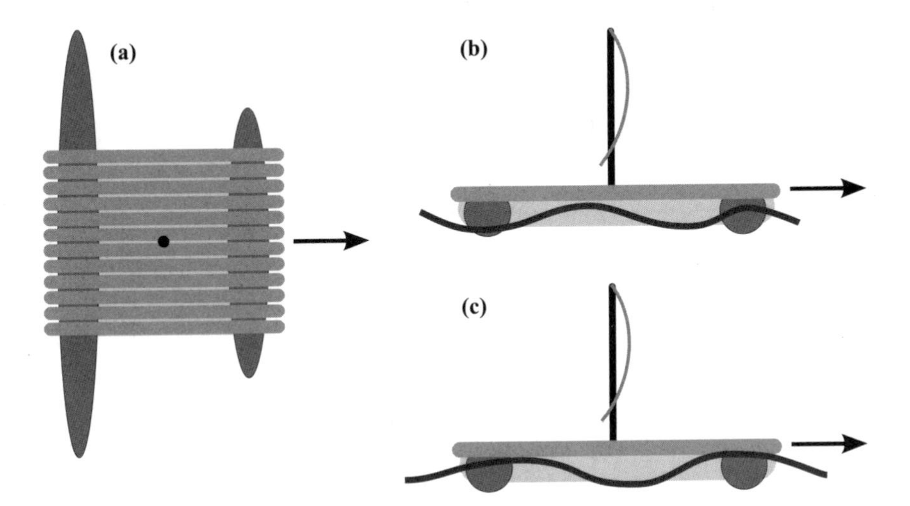

Figure 6.1. (a) Your hull-speed raft, viewed from above. Note the direction of motion. (b) When the bow wavelength is less than the distance between the long beams, drag is reduced compared to the case of (c). In (c) bow wavelength equals the distance between beams because the aft beam is more submerged. So hull speed is reached when hull length equals bow wavelength. Consequently, hull speed is limited by hull length.

force that is acting to hold back the boat increases rapidly once hull speed is reached. My goal in this section is to explain to you, in simple physics terms, why these phenomena occur.

Which is why I have press-ganged you into service onboard the undignified vessel illustrated in figure 6.1. She is a wooden raft with two long logs fore and aft that stretch way beyond her beam. These logs are not there to provide flotation, please note—we will suppose that the raft has enough buoyancy without them—but rather to illustrate hull speed. You set the primitive sail and drift off to the right. The forward log generates a bow wave which spreads out in the wake, as waves do. You notice something that you have seen many times before in other craft: the bow wave size (amplitude) increases as the vessel speed increases. This makes sense because the hull is pushing water aside, the displaced water has to go somewhere, and the faster you go, the more water is moved. So the wave size increases. Now you pick up speed, and so the wavelength of the wake, as observed alongside your hull, stretches out until exactly one wave lies between the two extended logs at bow and stern. The raft speed that gives rise to this condition is her top speed, you

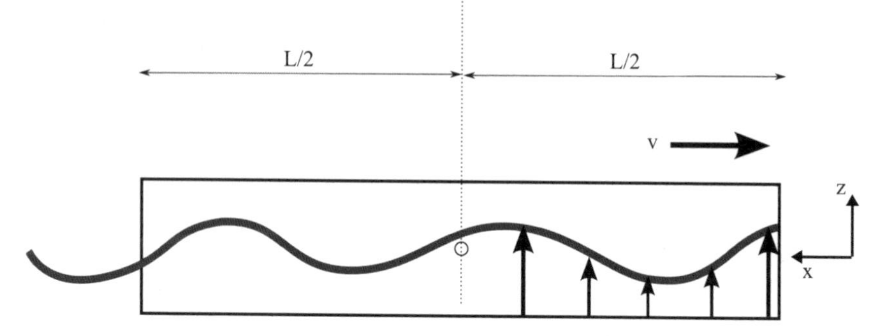

Figure 6.2. Your hull-speed barge. Bow waves forward of the center of gravity, CG (open circle) exert a buoyancy force (vertical arrows) proportional to wave height that acts to rotate the barge hull counterclockwise. Similarly, waves aft of the CG act to rotate the hull clockwise. If we can assume that drag forces are proportional to counterclockwise torque (a dominant CCW torque means that the barge is climbing a hill created by its bow wave), we can show that hull speed occurs when bow wavelength equals hull length.

find. It is clear why: the aft log is now submerged, and so experiences more drag than it did earlier, when there was no wave crest at the hull stern (see fig. 6.1). So, drag force peaks when bow wavelength equals hull length, in this simple example.

Now we are able to see where the old formula for hull speed comes from. The speed of a bow wave, or of any other surface water wave,[1] is c where $c^2 = g\lambda/2\pi$. Here λ is the water wavelength, and g is the constant acceleration due to gravity. Now the raft speed, v, equals the water wave speed, c, so that $v = \sqrt{gL/2\pi}$ (since hull length, L, equals water wavelength at hull speed, as we just saw). Substitute numbers and we arrive at the old formula.

The ungainly raft has served her purpose, and you can now abandon her. The lesson learned is intuitive, and yet it gives us a basis for understanding quantitatively what hull speed is about. Now I can do another calculation, this time a little more realistic. The math is more involved (you need not wade though it), but the basic idea is again quite intuitive. Figure 6.2 shows the profile of a steep-sided hull plowing through water and generating a bow wave, which oscillates along the line of the hull. This vessel is kept afloat by the buoyancy force, and we can see that the buoyancy force is going to be different at different points along the line of the hull because the wave height varies along the hull. Buoyancy that acts forward of the hull CG (shown in fig. 6.2) will create a counterclockwise torque that tends to twist the hull about the CG—trying to make it do a backflip. The buoyancy force aft of the CG produces a torque that acts in the clockwise sense. These two more or less cancel* but not quite. If the counterclockwise buoyancy torque is just a little bigger than the clockwise torque, the boat will tilt backwards, until her stern goes deep enough to generate a compensating torque. We would then be left with a boat that is going uphill, trying to reach the crest of her own bow wave.

Where am I going with all this? Roughly speaking, counterclockwise torque equates to uphill motion, and uphill motion leads to increased drag, for reasons that will soon be made clear. So, I am saying that increasing the unbalanced counterclockwise torque generated by a bow wave will increase drag. If this increase should suddenly take off at a certain speed, then we have found our hull speed. In fact, I can calculate the torque generated by the bow wave. You can see that as the bow wavelength changes, the torque will also change because the manner in which buoyancy force is distributed along the hull length changes with wavelength (fig. 6.2). The results of this calculation are plotted in figure 6.3. (For those interested, the math is provided in this endnote 2 in sufficient detail for you to reproduce the calculation.[2]) In figure 6.3 we see once again that drag force takes off for water wavelengths exceeding hull length, more or less.[3]

For simplicity, the hull of figure 6.2 was given vertical sides, but most boats don't have vertical sides, for a host of reasons. Recall that, in the Age of Sail, ships of the line were given a tumblehome cross section to deter boarders. Nowadays we are less likely to have to repel nefarious enemies swarming over our gunwales with cutlass in hand, casting a single bloodshot eye (the other being patched) in search of our gold doubloons. Hull sides are angled but the other way, with cross sections resembling a martini glass rather than a brandy glass. In plain language: more V-shaped. Here are some physics reasons for different hull cross sections.

- Rounded hull bottoms are stronger than V-shaped hulls, but the latter will be deeper for the same displacement and so will better resist leeway.

*Just as well, because backflipping boats would be pretty uncomfortable.

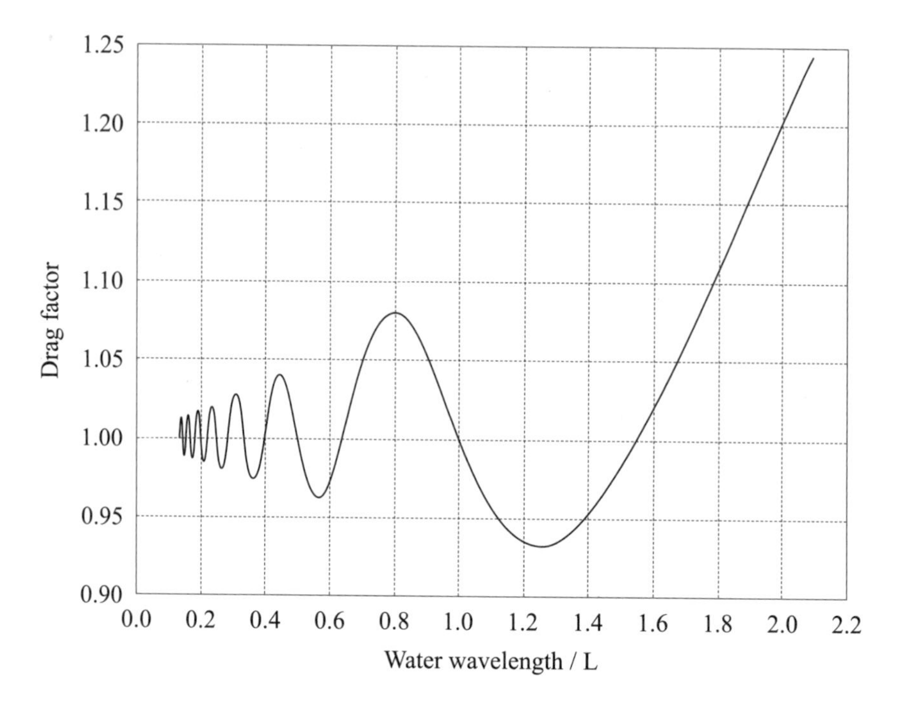

Figure 6.3. Hull speed is limited by drag. In the simple model described in the text, the drag increases with water wavelength, λ, as shown (L is hull waterline length). Here, drag force is set arbitrarily to 1 at zero speed. If the bow wave is assumed to have constant amplitude, independent of speed, then drag changes with speed as shown. For a more realistic model, with bow wave amplitude increasing with speed, the curve looks similar. In this simple model, hull speed occurs at $\lambda \approx 1.2L$ because for longer waves (higher boat speed) the drag force becomes too strong.

~ A large deck area is desirable, but large hydrodynamic drag is not. For a hull of a given displacement, the choice of hull shape is constrained by the trade-off between these two characteristics.

~ An angled hull—say one that is V-shaped—will have greater reserve buoyancy. That is, the righting moment will increase as the hull heels further and further.

~ During heeling, the waterline along an angled hull will not be symmetric about the longitudinal axis; the port side waterline length and shape will be different from that on the starboard side. This asymmetry can assist the boat to head up while heeling. Thus, even without aerodynamic assistance from her sails, a boat may automatically point to windward when heeling solely because of hydrodynamic forces acting on the hull.

- Different angled hull shapes beneath the waterline assist with planing. For certain boats, such as racers, this is important because planing requires less displacement, less wetted area, and so less drag—and hence increased speed.

The physics of angled hull shapes casts an interesting light on the capabilities of some ancient ships. Certain ancient ships were built with a lot of overhang at the bow and stern, but this practice is usually thought to have been of little value for the old square-riggers because these ships were supposed to be nippy only when running or on a broad reach. Today, such hull shapes are utilized to increase hull speed while heeling because the waterline length is increased when the hull is heeled over. This lengthened waterline increases boat speed on a beam reach, for example. It seems plausible to suppose that ancient vessels with overlapping bows and sterns may have been capable of traveling across the wind at speed. Indeed, such a hull design offers no other advantage for these square-rigged vessels. (An overhanging bow and stern increases deck area, but for merchantmen—and in ancient times most of the sailing ships were merchant vessels because warships were oar-powered—deck area was not such a big deal. Volume of the hold was what mattered.) For a downwind point of sail, extended hull length above the waterline will increase pitching motion when traveling downwind; this is bad, and yet the overhanging bow and stern must have conferred some advantage or these ancient ships would not have been built this way.

Keel Appeal

In chapters 1 and 3 we encountered the structural keel—the long structure that extended along most of the underside of a wooden boat and provided the foundation on which the boat was built. Here, we will examine the hydrodynamic keel—the underwater foil (hydrofoil) introduced in modern sailing vessels to improve sailing performance, in ways that I will now discuss.

The modern keel serves two or three separate functions that are not entirely compatible with each other. So, again, we are talking about design trade-offs. The functions of hydrodynamic keels are to

~reduce leeway;
~add ballast, and so reduce heeling; and
~perhaps provide vertical lift.

We saw in figure 5.1 how the keel is supposed to provide a sideways hydrodynamic lift that opposes aerodynamic drag, thus reducing the leeway made by a sailboat. In fact, perversely, a boat must drift—make leeway—in order for the keel to provide lift to resist leeway.* The reason is shown in figure 6.4. Because a keel must work equally for both port and starboard flows, it has to be symmetric about the longitudinal axis of the boat. It must, in other words, be aligned along the axis of the hull. Because of this, the keel provides no hydrodynamic lift when a boat is facing directly into the water flow. This is entirely analogous to a symmetric airfoil, which provides no lift unless it is cambered to the airflow. Consequently, some leeway is inevitable, though modern keels are very effective in minimizing leeway. For example, if the water-flow angle of attack is 5°, a typical keel will provide a hydrodynamic lift/drag ratio of 20:1. Hydrodynamic forces are significantly greater than aerodynamic forces for the simple reason that water is much denser than air, and so keels can be a lot smaller (with much less area presented to the flow) than sails. A small keel can easily provide enough hydrodynamic lift to match the aerodynamic drag of a large sail.

Of course, as with sails the shape of the keel is important in determining L/D ratio. As with sails, a long, thin foil is more efficient than a short, fat one. Consequently, the keels of racing yachts tend to be thin and long, like a glider wing on end. Why should a long, thin keel or sail—one with a high aspect ratio—perform better aerodynamically than a short, fat one of the same area? The aerodynamic reason is that vortices form at the tip of the keel, or sail, or wing, and this vortex formation (and shedding) reduces the lift force near the tip. A keel with a thin tip suffers less lift loss due to tip vortex formation and shedding than does a keel with a wider tip. A fuller explanation is postponed until the appendix; here, we need only grasp the fact that lift is reduced near the tip of a keel, and so keels with small tips provide greater lift.

This fact means that racing yachts have a deep draft and consequently have trouble in shallow harbors, and so we encounter our first keel trade-

*Unless there is a current that acts in a suitable direction.

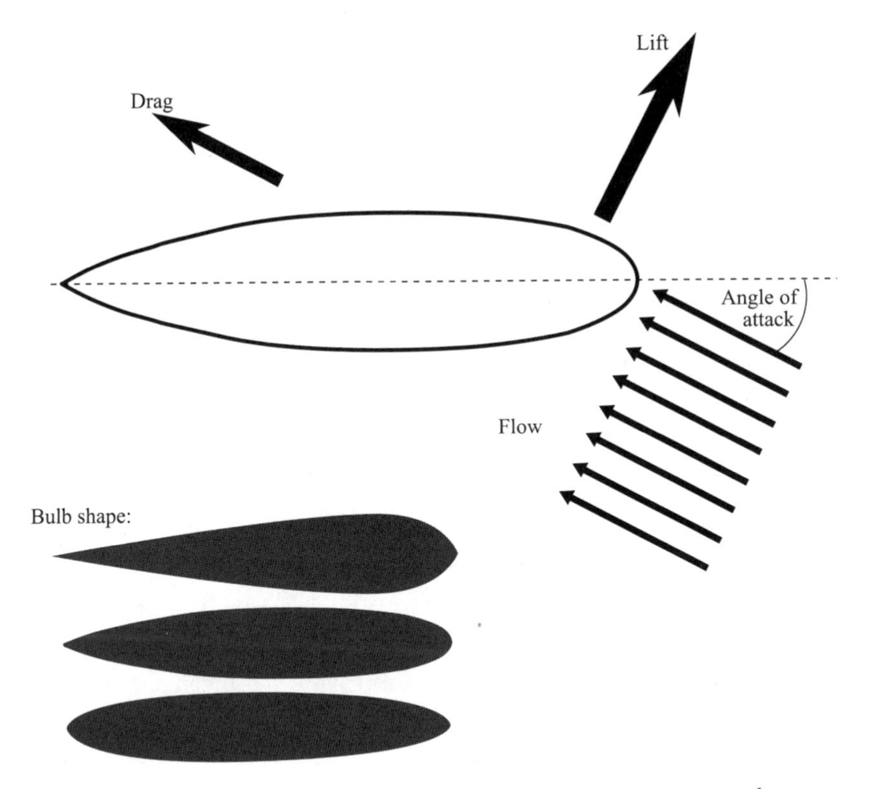

Figure 6.4. The keel cross section (at top) has to be symmetric to resist leeway motion from both sides. This means that it provides lift only when the flow is from the side. Keel bulb shape matters: of these bulb shapes, all of which have the same volume, which offers least drag?

off. In practice, keels are designed to perform as well as possible given the fact that many weekend sailors like to put into port from time to time. So keel depth is limited by considerations other than hydrodynamic lift efficiency. Another reason for wanting a deep keel is ballast. A longer keel means greater righting moment because the torque provided by a keel depends on keel length as well as weight. Thus, a deep keel means more hydrodynamic lift (and so less leeway) and also more righting moment (and so less heeling).

One common method of compromising between keel length and keel righting moment is to provide a bulb on the end of the keel. This adds weight where it is most effective in providing a righting moment, and so in resisting heeling. But bulbs increase the cross section of a keel that is presented to the water flow, and so we come to trade-off #2. What shape should the bulb be to maximize volume (and therefore weight) while minimizing hydrodynamic drag? Clearly this is a trade-off because increased volume means increased surface area, and increased surface area means increased drag. There are a number of different bulb shapes out there—testimony to the effort that has gone into designing the most effective shape. One factor that seems to have emerged is that the most efficient shape results when the maximum keel bulb width is 45% of the distance from keel bulb front to aft, as illustrated by the middle of the three bulb shapes in figure 6.4.

Another attachment on the end of some keels is a small hydrodynamic wing. The primary reason for such winglets—on boat keels as well as airplane wingtips—is to interrupt the formation of tip vortices and reduce the amount of energy that such vortices bleed from the moving vessel as they shed, resulting in less induced drag. (Given the complexity of vortex physics, a fuller explanation is relegated to the appendix.) A secondary purpose of this winglet is to provide vertical lift, thus raising the hull and reducing the hull wetted area, and so reducing hydrodynamic drag. A small keel wing can provide significant lift—much greater than an airfoil of the same size—because water is much denser than air. Careful design is necessary here, too, because (for example) the lift may not always be vertical if the hull is heeling. The implication is that the hull righting moment depends on speed through the water (because the keel wing lift will depend on speed)—yet another example of design constraint, as one aspect of boat performance impacts on another.

A different compromise, one that provides an effective deep keel in deep water and shallow keel in harbor, is the retractable keel—a centerboard or daggerboard. Retractable keels present technical problems and take up space in the hull: this is the compromise. Suitable for small sailboats, the centerboard is a movable (pivoted) keel that is stowed in a slot in the centerline of the hull (often the structural keel). It is lowered when needed to resist heeling or leeway, and is raised for shallow draft or to lower drag when the boat is running with the wind. As with the fixed fin keels, the cross section has to be symmetric about the boat's centerline. When stowed, the centerboard swings up and aft. Thus, the pivot can be used to *partially* raise the centerboard in order to move the CLR aft if the sail plan calls for this.

The daggerboard performs the same functions as the centerboard but is raised vertically rather than pivoted. Thus, the daggerboard cannot be used to shift CLR. In fact, the primary use of daggerboards is more to reduce leeway than to perform any of the other keel functions because it is usually longer, thinner, and lighter than a centerboard.

Two other retractable keels are bilgeboards and leeboards. These come in pairs, mounted symmetrically on the port and starboard sides.* The advantage of two foils, rather than one, is that they can be asymmetric (canted to the current, or constructed with significant camber), thus providing superior lift. This works because when the craft is heeling, the windward foil is raised out of the water and the leeward foil is more nearly vertical.

Keels are not designed in isolation and then fitted willy-nilly to any boat or ship hull, of course. There is significant interplay between hull and keel; they must be carefully matched so that the boat handles well. Because hull shape conditions the flow of water past the keel, hull shape affects keel performance and vice versa. Different combinations of hull and keel produce boats with different characteristics. A long, narrow, shallow hull with no keel will run with the wind very fast (little induced drag) but will capsize easily (little righting torque). A deep, heavy keel will make a boat practically uncapsizable but will add drag. A light, shallow-draft flat-bottomed boat with a centerboard is a reasonable compromise in sheltered waters but will be uncomfortable in open waters (too much righting torque) and difficult to handle (not much momentum—she will lose headway when tacking). Thin keels will stall at lower speeds than fat ones† but produce less drag. Long keels stay on course better than short ones but produce more drag. They are better to windward and are better at resisting heeling but require deep water. Bulb keels also provide more stability and so permit more canvas. Winged keels reduce hydrodynamic drag and provide lift as well as ballast, but as more than one observer has wryly noted, they are also better at catching kelp.

The message that comes across from this brief survey of keel physics is that most boats are a compromise on keel and hull dimensions; the keel is matched to the hull in different ways depending on the desired capabilities of the boat.

* Leeboards are mounted on the sides of the hull, and bilgeboards between the sides and center.

† The phenomenon of stalling is briefly aired in the appendix.

Rudders

It is entirely possible to sail over long distances in a rudderless boat. We have seen that torque can be applied to change heading by shifting CE and CLR—for example, by trimming sails or shifting crew weight around the deck—and allowing the wind to do the rest. However, fine course correction is made much easier by the use of a stern rudder, literally a pivotal invention that we encountered in chapter 1. Rudders fall into two broad categories: outboard and inboard. Outboard rudders, hung on the stern or transom, are common on small boats. Larger vessels may have an inboard rudder, which is hung from the keel and (unlike the outboard rudder) always fully submerged. The inboard rudder is connected to steering (a tiller or wheel) via the rudder post.*

The rudder, like the keel, is a hydrofoil, though one with different characteristics and with different demands placed on it. It must generate hydrodynamic lift, but not just along the boat's heading direction. The rudder must produce lift for a larger range of angle of attack than the keel (typically between about 3° and 10°) and has to operate in the turbulent water of the keel's wake. We have seen that most boats have an intentional imbalance between CE and CLR that usually leads to a weather helm—the boat naturally tends to turn upwind. The job of the rudder is to correct this tendency and maintain course. So, for example, if *Puddleduck* carries a weather helm, then you will hold the tiller to the windward side so that the rudder projects leeward of the keel, to counter the weather helm. In harbor a rudder permits a drastic change of course over a small distance. Steering is accomplished by moving the tiller to port or to starboard, resulting in the stern of the boat shifting sideways, to starboard or to port. This sideways motion is referred to as *yaw* and can occur unintentionally (for example, as a result of changing currents) as well as intentionally.

We can best picture the action of the rudder by appealing to the simple momentum flux view of lift, as in chapter 2. Because the force provided by the rudder is hydrodynamic lift, the boat must be moving through the water in order for the rudder to be effective. Water strikes the front surface of the rudder and transfers momentum, causing the

*Recall that the difference between a ketch and a yawl depends on where the mizzen mast is positioned relative to the rudder post.

boat stern to shift sideways. A more sophisticated view of rudder lift is required when we look at the physics in more detail. It turns out that the foil shape of the rudder is more important that its size in determining steering effectiveness. Of course, as with the keel, the rudder cannot have any camber—it must be symmetric about its longitudinal axis—so that it can work on both sides.

Wake Up

Everybody is familiar with the idea of a ship's wake. We generally cannot see a wake very clearly on the sea because much of it is masked by other waves, but we certainly feel it in a small boat when a large wake passes underneath. On calm lakes and slow rivers, we see wakes in all their glory, and some of us marvel at the complexity and consistency of the wake wave patterns.* Physicists are fascinated by this sort of thing, and the scientific explanation of wakes is complex and interesting. The complexity and consistency of wakes is shown in figure 6.5. What do I mean by wake "consistency"? There are certain features of a wake that are constant, utterly unvarying, and remarkably precise. Thus, the *opening angle* of a wake—the angle of the V's in figure 6.5—is known to be 38.94°. This value was first calculated in the 1880s by the Scottish physicist William Thomson (elevated to the peerage as Lord Kelvin), and the V-shape that outlines all wakes is known to physicists as the *Kelvin envelope*.

The striking feature about the opening angle is that it is the same for every boat no matter how large or small it may be, how fast it is moving, or what shape it might take. In fact, the opening angle for the wake formed by a duck on a pond is the same as that formed by a supertanker on the ocean. Furthermore, the opening angle does not depend on the wavelength of the waves produced, or on their amplitude, or even on the liquid on which the wake is formed. You could sail *Puddleduck* on a lake of Scotch malt whisky (now there is a thought—Lord Kelvin would be appalled), and her wake envelope angle would still be 38.94°. The wake is formed by the plowing action of the boat (or the duck) through the water, which then propagates waves because of the influence of gravity—the

*By now, I hardly need point out that the physics of wakes is complicated: almost all of fluid dynamics is complicated. As usual, my explanation is qualitative in the text, where lucidity is the goal, and more detailed but mathematical in the endnotes.

Figure 6.5. The wake generated by a boat is distinctive, despite the universal nature of certain features. Images from Wikipedia.

manner in which water waves disperse determines the opening angle.[4] But even gravitational strength does not matter, so long as it exists: you could sail a brick on a lake of mercury on a distant planet, and the Kelvin envelope angle of your wake would still be 38.94°.

This consistency is a godsend to the naval and customs surveillance community. There are lots of spies in the skies looking down on the world's oceans, searching for ships and boats belonging to enemies or smugglers. From the height of a satellite, even the largest ship is hard to see amid the vastness of the oceans, but wakes show up more readily (see fig. 6.6). They are strung out for a hundred miles in some cases and are always much bigger than the vessel that gives rise to them. The consistency of the Kelvin envelope angle helps in the detection of wakes, from which the presence of a ship can be inferred. Clever detection algorithms process the remote sensing data of these surveillance satellites, looking

Figure 6.6. Wakes are much bigger than the objects that generate them and can be seen from space. Indeed, remote sensors such as military satellite imaging radar systems utilize wake detection algorithms to detect the presence of ships. In this case, however, the objects generating the wakes are islands in the South Atlantic Ocean; the wakes are caused by strong winds and are formed in clouds. These clouds are known, predictably, as "ship-wave-shaped clouds." Image courtesy of Jacques Descloitres, MODIS Rapid Response Team, NASA/GSFC.

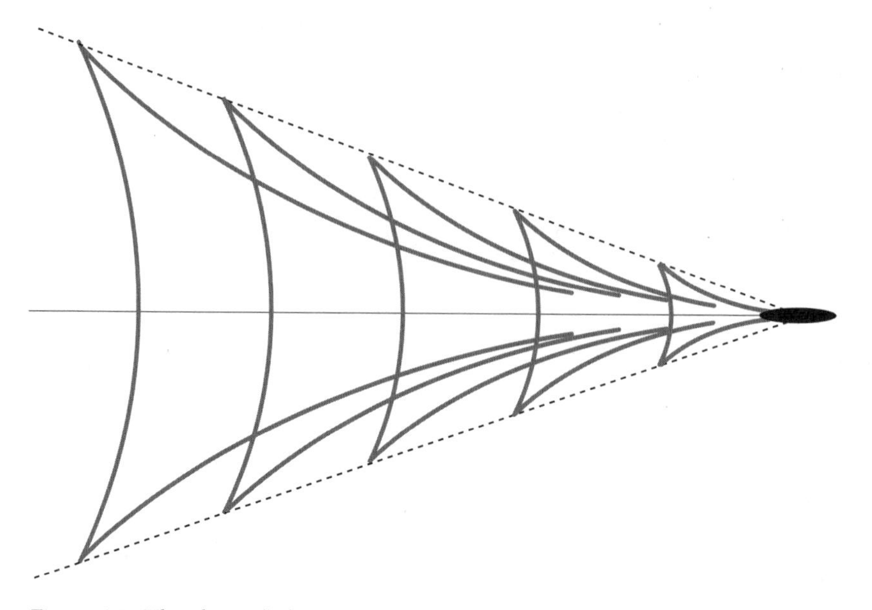

Figure 6.7. The classic "Christmas tree" wake pattern of a moving boat. Dashed lines show the Kelvin envelope, which forms an opening angle of 38.94°. Two types of waves form the Christmas tree: transverse waves, which are perpendicular to the boat direction of motion, and divergent bow waves, which all form an angle of 35.26° with the envelope.

for patterns of the Kelvin type. Every pattern that does not display the exact angle required is rejected, and so wake signals can be dug out of very noisy data. Once identified, the details of the individual wake yield information about the ship that generated it: speed, size, form.

But if the Kelvin envelope is consistent, and independent of boat speed and shape, etc., what can looking at a wake tell us about the boat? Well, the envelope is certainly the same for all boats, but other wake features depend on the boat that generates it. So, once picked out of the noisy data and identified, a wake becomes an individual signature, almost like a fingerprint.*

There are two main components of every wake and a host of minor components that vary from wake to wake. The main components are

*I exaggerate here. It may well be that wakes are as individual as fingerprints and that in principle they are a unique signature of a boat, but in practice individual boats cannot be identified from their wakes. That would be far too complicated a task in the real world. From the wake it produces, we *can* tell about the heading and speed of a boat, and can learn something of its form, as we will see.

two types of standing waves that interfere to produce the classic wake pattern shown in figure 6.7. Standing waves are waves that do not change: they literally appear to be standing still. For this reason they are easy to pick out amid a mass of other waves moving every which way. To see a standing wave, start your car engine and place a cup of coffee on the hood; a pattern of circles that does not dissipate is set up on the surface of your coffee. You can create standing waves in a bath by moving your hands at a certain frequency, like clapping in slow motion. The sounds that are produced in an organ pipe or when you blow across the top of a tube are acoustical standing waves. The standing waves that constitute a wake are more complicated than the examples that I have given but are of the same ilk. They are more complicated because they appear to be standing waves from the point of view of somebody on the boat; viewed from the shore, the wake waves move.

The two types of wave that emerge as standing waves in a wake are called *transverse waves* and *divergent bow waves* (fig. 6.7). Another example of constancy in ship wakes is the angle made by the divergent bow wave with the Kelvin envelope: it is always 35.26°. Those of you who are interested enough to plow through the math will find details of the derivation of wake standing waves in endnote 5.[5,*] In addition to standing waves there are other types of waves produced in wakes. Some ships generate a significant stern wave and, if the transom is large enough, one stern wave from each corner. There is a turbulent wake immediately aft of the stern which looks remarkably calm and remains so even for some considerable distance behind the ship. If a ship has propellers, these generate underwater wakes that percolate to the surface and add further "fine-structure" details to the observed wake. It is these extra details, not the constant standing waves produced by transverse and divergent bow waves, that yield information about the ship.

Apart from looking beautiful, wakes are a drag. To be more precise, they are a significant component of the hydrodynamic drag force—at least half. This makes sense: think of the energy required to turn aside all the water that makes up a wake. Given that wakes can persist for a long way behind a boat, it is reasonable to suppose that bow waves carry away a lot of kinetic energy. Much effort has been expended in shaping ship

*Those of you who are not into math but would like more information about wakes will find a good technical but math-lite account in Jearl Walker's article.

hulls so that they generate minimum wake. Recall the rounded sterns of the old clipper ships—such sterns generate less stern wake than the older and easier-to-build square sterns. Similarly and more intuitively, sharp and angled bows cut through the water more cleanly and may lift the bow somewhat to reduce wakes and so reduce hydrodynamic drag. You may have noticed waterline bulges on the bows of large ocean-going supertankers; these are designed to minimize bow wave generation, reducing both fuel costs and shoreline erosion.

In fighting trim: *Ready for battle.* Some warships of the late nineteenth and early twentieth centuries were able to reduce their freeboard by flooding water tanks, thus lowering their profile prior to combat.

Know the ropes: *To understand how an organization works.* The miles of cordage in an old square-rigger had many different names related to multifarious functions. An experienced sailor would know the ropes. Some sources indicate the opposite meaning: an inexperienced seaman's discharge may have written on it "he knows the ropes," indicating that he knew little else.

Tide over: *To support through a difficult period.* A ship run aground may await a high tide to free itself. A sailing ship stuck in port due to adverse tide and wind would await more favorable conditions for departure.

Under the weather: *Unwell, run down.* A sailor's watch on the weather side of a ship would expose him to wind, rain, and spray.

7

Windsurfing

Catwalk: *Narrow, elevated walkway.* Originally a raised bridge amidships, running fore and aft, providing a safe passage over deck obstacles.
Footloose: *Free to travel.* An unsecured foot line permits the lower edge of a sail (the foot) to flap freely.
Filibuster: *To stall or obstruct the passage of legislation by long speeches.* From the Dutch *vrybuiter* (freebooter or pirate). Pirates would stop sailing vessels in order to search and rob them.
Stranded: *Left in an inconvenient place.* A ship that had run aground on a beach (strand) was said to be stranded.

Sailboards

The equipment is called a sailboard; the action is better described as windsurfing. Some of you may argue that sailboards are not boats and so don't belong in this book. To forestall such mutinous mutterings, I note first that a sailboard is a hi-tech raft, that it is equipped with a fore-and-aft sail, and that windsurfing is hugely popular. Second, windsurfing brings out a few interesting aspects of hydrodynamics and the physics of sailing that are not apparent in sailing with boats.

Windsurfing is similar to sailing a boat in that the drive originates from the wind and is delivered to the hull (the board) via a sail. However, there are four key differences between sailing a boat and sailing a board. First, the sailor's weight is usually much greater than that of the board. A

Figure 7.1. A windsurfer trapezes to counter heeling torque. Thanks to Stefania Bocheva for providing this photo.

modern racing board typically has a mass of 5–7 kg (a weight of 11–15 lb), while a beginner's board is in the region of 8–15 kg (18–33 lb). Nowadays boards are made of polyethylene filled with PVC foam or of a polystyrene foam core reinforced with a carbon fiber/Kevlar/fiberglass shell. The density is such that the board's buoyancy is a close match to the weight of the board plus that of the sailor (the board does not sink much when the sailor clambers aboard). The fact that the sailor's weight dominates means that he or she can shift the sailboard's center of gravity at will and can trapeze (hang out of the sides, to counter a heeling torque) very effectively (fig. 7.1). The sailor can throw her weight around and do things that are unthinkable on a boat—for example, launching herself skyward or rocking the board to gain forward momentum (of which more later), and *ooching* (a forward body movement stopped abruptly, providing forward board momentum).*

The second difference is that sailboards are designed to plane. Some early, heavy boards were displacement hulls, but today, sailboards are almost always planing boards. So we expect that sailboards when planing will be subjected to much less hydrodynamic drag than are yacht hulls.

The third difference is that boards have much less resistance to leeway drift than do most yacht hulls. Clearly, this is related to the flat-bottomed planing design. Additionally, sailboards have no keel, though beginner's boards are often fitted with a small retractable daggerboard. The emphasis in sailboard design is clearly placed on drag reduction, and in any case leeway is not the problem that it is for yachts, where heading direction and precision steering within harbors is essential. Windsurfers steer by manhandling the sail and by shifting their weight around the board. The facts of reduced drag and increased leeway must be taken into account when we come to examine the equation of motion of sailboards.

The fourth difference is that windsurfing takes place in shallow water, whereas yachting usually doesn't. To a hydrodynamicist, there is a great difference between deep water and shallow water, having to do with the behavior of waves on the surface. In the next section we will see how water depth influences waves and how the different characteristics of shallow water influence the physics of windsurfing.

In addition to these four key differences, there are differences of scale. Because sailboards are smaller than most boats, they are more susceptible to small gusts of wind and react more quickly to a sudden change in wind direction. Sailboards are smaller than many waves, and so waves influence sailboard motion significantly—indeed, some would say, this is the whole point (figs. 7.2 and 7.3).

Before proceeding to the physics I need to provide a little more background information on windsurfing and sailboards to show what motivates windsurfers and to provide a backdrop for the calculations that follow. The sport of windsurfing developed quickly in the late 1960s and

*Ooching works only because of drag, or friction, between the board and the sea. Without drag, Newton's laws tell us that no net momentum gain can accrue by the sailor's throwing herself around her board, however hard she tries.

Figure 7.2. A windsurfer launches his board skyward from a small wave. Image courtesy of CoastalBC.com.

has exploded over the past four decades to become a hugely popular recreation and Olympic sport. Given the popularity of windsurfing and the multimillion-dollar industry that it has given rise to in various parts of the world, you may not be surprised to hear that there is acrimonious, litigious disagreement over who first invented the sailboard and when. I have no intention of going there: all we need to know is that sailboards are a clever combination of surfing board and triangular sail, wedded together successfully because of the strength and lightness of modern materials.

Originally, boards were classified as being either short or long, with shortboards being less than 3 m (10 ft) in length and designed for planing. Longboards exceeded 3 m in length; they were equipped with a daggerboard and were intended for racing. This classification is now obsolete. Modern boards tend to be shorter (2.3–2.6 m, or 7½–8½ ft), and almost all of them plane. They are faster and more maneuverable and have become specialized for different purposes.* Slalom boards were developed for top speed. Beginner's boards are more stable and easier to handle, and waveboards are for use in breaking waves. There are freestyle

* Though fun seems to be the main purpose for all sailboard designs.

boards, wider and with a greater volume than other boards, designed for performing acrobatic tricks such as jumps, rotations, slides, flips, and loops. Other styles of boards also exist. All these modern board styles are classified by board width and volume (rather than by length) into one of three broad categories. *Formula* category boards are racers with large sails (up to about 12 m^2); *wave* category boards are shorter, technically demanding, and intended for stunts and big waves; *freestyle* category are all-rounders.

The sails are specialized also, in size and shape. The sail is made of Dacron or Mylar, with Kevlar reinforcement, and is attached to a strong but flexible carbon mast and boom (fig. 7.4). Sail area varies considerably, from the racers' 12 m^2 down to 3 or 4 m^2 for waveboards. Racing sails are cambered (curved, and held in shape by battens), while wave sails are reinforced and flat. Freestyle sails are in between, sacrificing speed or stunt performance for comfort and (relative) ease. There is no rudder, and steering is performed by the sailor, with his or her back to the wind, holding the boom to maintain or alter the sail angle of attack. Consequently, the drive force goes through the sailor to the board. Windsurfing happens on flat seas or breaking waves, and in light breezes or stiff winds. The balanced combination of board and sail—and sailor—leads

Figure 7.3. A windsurfer launches his board skyward from a large wave. Thanks to Rich Swanner for permission to reproduce this photo.

to impressive sailing performance. As we will see, the speed record for a sailing craft on water (though not on ice) is currently held by a windsurfer, not a yachtsman.

Breaking Waves

The equation for the speed, c, of a surface wave across the ocean is well known to physicists and is rather complicated. In deep water this equation reduces to a simple form that we have already encountered, $c = \sqrt{g\lambda/2\pi}$. So, in deep water the speed of a wave depends on wavelength, λ. This fact influences the hull speed of a ship, as we have seen. In shallow water, the complicated equation for c reduces to a different limit, $c = \sqrt{gh}$, where h is the water depth. So, in shallow water the wave speed does not depend on wavelength. Water is considered to be shallow if the depth is less than half the wavelength of the wave. Thus, if two waves are moving over the surface of a harbor that is 20 m deep, and the first wave has a wavelength of 100 m while the second has a wavelength of 15 m, the harbor appears shallow to the first wave and deep to the second.

Figure 7.4. Hi-tech surfing. The board, boom, flexible mast, and sail are all made from modern materials. Image from Wikipedia.

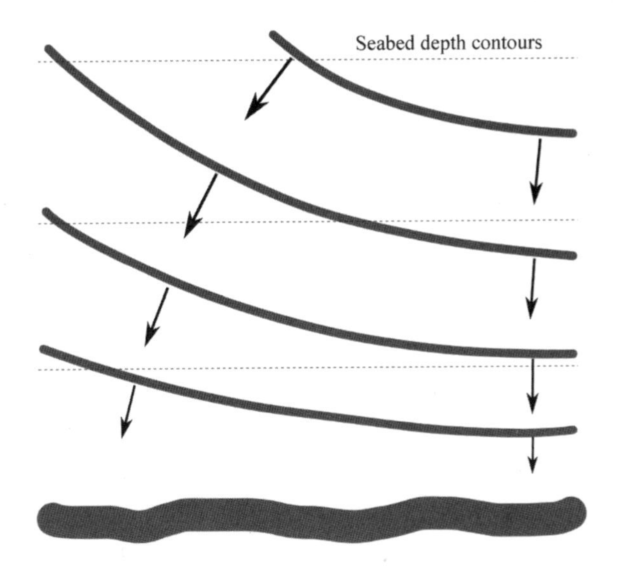

Figure 7.5. If a wave approaches a shore obliquely, as seen here from a bird's eye view, a section of wave near the shore is in shallower water than a section further out and moves more slowly. So the wave changes direction to approach the shore less obliquely, i.e., more nearly parallel to the shore. In addition, the wavelength (the distance between the crests, shown here) decreases as the waves approach the shore.

There are several significant consequences of the shallow water limit for wave speed. Note that the wave slows down as the water becomes shallower. As a wave approaches a gently shelving beach, its speed slows (to zero, at $h=0$). The first consequence of this fact is that waves approaching a beach "shape up" to the shoreline; that is, they change direction so that they become more parallel with the shore.[1] You can see how this works in figure 7.5. Wave speed, $\sqrt{gh}$, decreases as a wave approaches the shore because h becomes less and less as it nears the shore; a wave approaching the shore at an oblique angle will straighten up.

The second consequence follows from the first. Again from figure 7.5 we see that because waves slow down as they approach a shore, they bunch together. But the energy of a wave has not dissipated significantly, and so the wave energy is contained in a smaller and smaller space as the waves bunch closer and closer together. This results in an increase in the wave amplitude. Think of water as an incompressible liquid (which is very nearly true). As the waves approach the shore, they become shorter. Because their volume does not change, they must grow taller. These

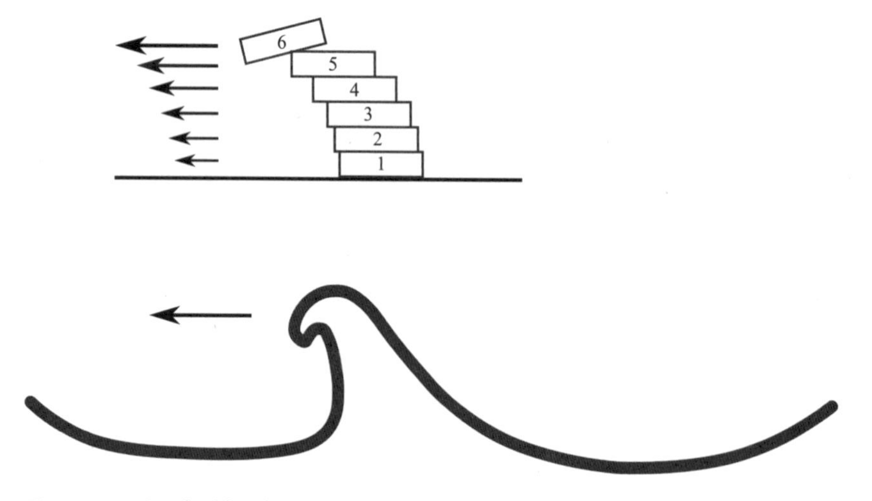

Figure 7.6. Stacked books on a conveyor belt will slip as shown if the conveyor slows down abruptly. Book 1 slows down almost as much as the conveyor because it has the weight of five books above it to hold it in place. Book 2 retains a little more of its speed because there is less weight above holding it down and more opportunity to slip (slippage occurs between conveyor and book 1, and also between books 1 and 2). Book 3 slips even more—i.e., retains more of its speed—and so on. Book 6 benefits from the slippage of all the books beneath it and so is least influenced by the conveyor deceleration. Breaking waves are analogous to the stack of books.

consequences of wave speed are familiar: as waves approach a beach they bunch together, shape up to the shoreline, and grow taller.

The physics of water waves backs up these everyday observations. We can show mathematically why the wave speed decreases in shallower water and how this slowing leads to increases in wave amplitude, etc. However, you have just seen that the underlying physics can be explained without math. A gradually shelving beach has another consequence for water waves that can be explained mathematically only with great difficulty but can be explained in physics terms quite easily, and that is the phenomenon of breaking waves. As waves slow and grow, they tip forward and break. Usually this wave breaking occurs when the amplitude reaches about three-quarters of the water depth (the surf zone of a beach). We can understand why this breaking phenomenon occurs in a couple of ways. First, from the equation for wave speed in shallow water ($c = \sqrt{gh}$) we see that the crest of a wave will move faster than the trough

Figure 7.7. A spilling wave. Image from Wikipedia.

because *h* is greater at the top than at the bottom. The top of a wave gets ahead of the water beneath it, as shown in figure 7.6, and the wave breaks. Physically, we can see what is going on by imagining the wave to be like a stack of books on a conveyor belt. If the conveyor belt is suddenly slowed down, the books will continue trying to move forward, due to inertia. The result is shown in figure 7.6.

Oceanographers, hydrodynamicists, shore erosion ecologists, and all those intelligent people who are about to read this paragraph recognize four different types of breaking waves, named for the manner in which the wave crest collapses: *spilling, plunging, surging,* and *collapsing*. (Surfers cast expert eyes on waves that approach the shore, of course, but they are looking for different wave characteristics than are scientists, and so they give waves, and wave characteristics, a different set of names.*) Spilling waves are white-tops, generating a frothy top as the crest breaks up (fig. 7.7). The crests of plunging waves remain intact; they turn over (perhaps forming a "tube") and then crash dramatically; these are the surfer's waves. Surging waves occur on steeply sloping beaches; the base

* Beach breaks, corduroy, crumble, glassy, ground swells, point breaks, tubes . . .

of the wave reflects off the beach and back out to sea before the crest breaks properly. Collapsing waves are somewhere between plungers and surgers. The last two types are of no interest to surfers.

We have seen how the equation for wave speed in shallow water greatly influences the behavior and appearance of waves as they approach a beach. There is another physical effect of shallow water that is not so apparent if you are just looking at the waves but which is nevertheless of great practical importance. In figure 7.8 we have two waves merrily progressing from right to left. The deep-water wave maintains constant wave amplitude and wavelength. The wave approaching shore behaves differently, as we have seen. What about the individual water molecules? They do not move along with the wave but instead circulate as shown in figure 7.8. This is a surprising revelation to some people, who expect that, somehow, the water molecules just follow the wave, but this is not generally the case.* It cannot be so, if you think about it, because if the water moved in the wave direction, there would soon be no water left at the source of the wave disturbance.

Water molecule circulation becomes flattened near the sea bed, and in shallow water the flattened ellipses become stretched out almost to straight lines. This behavior of the individual particles of water is well understood mathematically and has been verified experimentally on many occasions. The fact that shallow water particles move almost horizontally, and over large distances, makes beaching waves dangerous. A tidal wave is a wave defined by this type of behavior. Tidal waves are dangerous because the water particles do not move in little circles or ellipses, but as a body in the wave direction when the wave passes by, and then as a body in the opposite direction as the wave is sucked back into the ocean.

Tsunamis are extreme examples of this type of wave. They originate in deep water, perhaps thousands of miles from shore, as very long wavelength waves with a lot of energy due usually to an underwater earthquake. The waves are so long (around 100 miles—160 km—in wavelength) that their enormous energy is not apparent out at sea. They will have low amplitude and may pass beneath a ship without anybody on

*Perhaps this misconception arises from looking at waves that break on shore; the molecules in these waves do indeed move along with the wave (or what is left of the wave) as it hits the beach, as we will see.

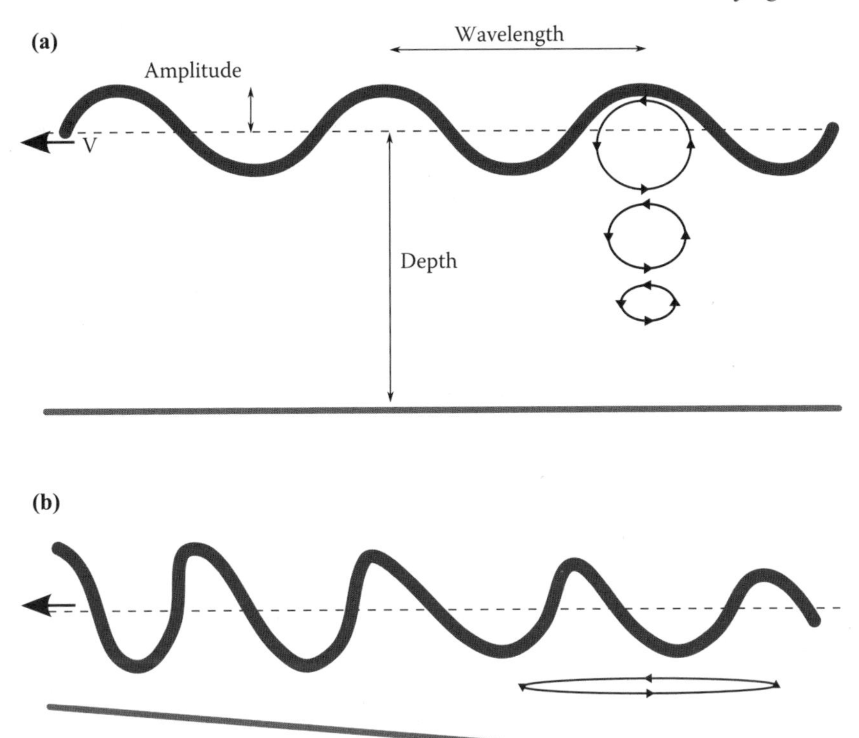

Figure 7.8. (a) A wave in deep water does not change shape: the crest amplitude and the distance between crests (the wavelength) are pretty much constant. Individual water particles move through circles or ellipses (in this case counter-clockwise, since the wave is moving from right to left) as the wave passes over them. (b) Waves approaching a shore act very differently. Their amplitude increases and wavelength decreases. The front face of the wave becomes steeper and eventually breaks. Individual water particles describe elongated ellipses (more like straight lines), and they can move considerable distances from their equilibrium positions as the wave passes by.

board even noticing. But as they approach a shore, these waves become shorter and higher. Much higher. As shallow water waves, they move onto the shore as a wall of water, with all the water particles moving together, and then all receding together. This, of course, can devastate coastal towns and villages, as we all saw over the Christmas vacation in 2004, when a large tsunami in the Indian Ocean came ashore and took the lives of over 150,000 people, mostly in Sumatra. The rapidly approaching wall of water caused great damage, of course, but so did the receding water. The volume of each tsunami wave is very large. For example, we now know that the wave trough that preceded the deadly waves in the 2004 tsunami caused the shoreline to recede 150 m (500 ft) just before the wave's arrival.

So, from figure 7.8 we see that those water particles which constitute a shallow water wave move along with the wave as it hits the beach. Most waves are not as damaging as tsunamis because they possess incomparably less energy, but they all behave in the same way. For windsurfers the behavior of shallow water waves is important because they are floating on these waves and are carried along by them. I could realistically ignore water speed when calculating the motion of sailing vessels in open ocean, but I will have to take it into account when considering how sailboards move in shallow water.

Planing and the Need for Speed

Surfers require waves in order to move, and windsurfers certainly will hitch a ride on one if they feel like it, but windsurfers can also achieve speed on flat water. The current speed record for any sail-powered craft on water is held by windsurfer Finian Maynard of the British Virgin Islands, who maintained an average speed of 48.7 knots (54 mph) over a 500-m course in April 2005. This achievement beat the previous record of 46.5 knots set by a catamaran in 1993. The nautical mile record is also currently held by Maynard (39.97 knots), according to the World Sailing Speed Record Council; Maynard's record has recently been beaten by Bjorn Dunkerbeck, who reached 41.14 knots, though this speed has yet to be verified. Clearly, over long distances yachts are always going to beat sailboards, and the yacht speed specialists are catamarans.* Equally clearly, over short distances the sprint specialists are now sailboards. The records are falling quickly, and I would be willing to bet that the 500-m and nautical-mile records just quoted will be out of date by the time this book goes to print.

The big advantage that sailboards hold over yachts is that they can plane over the water rather than plow through it (see fig. 7.9; planing is

*You are now in a position to understand why: cats have less wetted surface and so less drag than monohulls. Less drag means more speed. Catamarans have a wider beam and hence greater initial stability, which means that they can fly taller sails without a likelihood of capsizing. More sails mean more drive means more speed.

Figure 7.9. By standing well aft and building speed, this windsurfer has caused his board to plane, thus reducing drag and increasing speed. Image courtesy of CoastalBC.com.

also evident in figs. 7.1 and 7.4). So how does a sailboard permit planing? It is a simple hydrofoil—a wing that provides vertical lift by presenting itself to the water surface at a controllable angle of attack. This is a particularly simple type of lift to visualize; as with, for example, a water ski or a rudder, we can think in terms of momentum transfer. Water impinging on the lower surface of the board is deflected downward so that the board is pushed up. A modern sailboard with a 6–8 m^2 sail will plane when it reaches a speed of 12 knots. If the sail area is increased to 10–12 m^2, the minimum planing speed is reduced to 8 knots. Wide boards plane at lower speeds; they do not reach the top speeds of narrow boards because of their greater drag. Boards with a tail rocker—an upward curve at the back, which provides improved maneuverability—require higher speeds before they plane. There is a hull speed barrier to overcome before planing is attained, but this barrier is small for sailboards.

The physics of chapter 4, in which I developed a fairly simple equation of motion for fore-and-aft rigged vessels, can be taken over (with suitable adjustments) to describe sailboard movement. The parenthetic "with suitable adjustments" requires some explanation because of the significant differences between Bermuda-rigged yachts and sailboards. Obviously, parameters such as mass and sail area change from yacht values to sailboard values, but the differences go deeper, as we have seen. That is, I cannot just apply the equations derived in chapter 4 with windsurfing parameters substituted for yacht parameters. I must account for these facts: (1) sailboards make significant leeway—they are blown downwind because they lack a keel to resist such motion; and (2) sailboards operate in shallow water.

The island of Tiree off the west coast of Scotland is isolated, barren, beautiful, and full of windsurfers. Those of you who have been to Scotland will know that the climate there is not exactly the balmy, sun-drenched, palm-fringed paradise that windsurfers find in Hawaii, so what is it about this particular island that draws windsurfers on the 8-hour ferry ride from the mainland of Scotland? Like many of the Western Isles, Tiree is buffeted by winds during most of the short summer months, but unlike other islands, Tiree has beaches all around its coast. Consequently, it is always possible to find a beach for which the capricious winds blow onshore. This is important for windsurfers because if they drift on the ocean surface for whatever reason, they would like to drift toward a beach rather than out to sea. In more benign climates with more reliable winds, they may be reasonably sure that a chosen beach will always provide them with a safe onshore breeze (and big plunging waves, as in Hawaii), but the weather in Scotland is harsh and variable—and so Scottish windsurfers go to Tiree.

This last paragraph was not inserted at the behest of the Tiree Tourist Board but instead serves to show that, quite often, windsurfing is carried out on beaches with an onshore breeze. I will assume in my calculations that this is the case. Thus, the wind direction is pretty much the same as the wave direction—recall that waves straighten up when approaching a beach. So, waves will push our sailboard towards the beach, and wind will cause it to drift in the same direction. I can simply add these two effects when accounting for them in my physics calculations. This is most conveniently done by assuming that the effect of wind and water together cause the sailboard to drift with the wind at a speed sw. Recall that w is wind speed, so here I am saying that leeway and wave drift combine to cause the board to move with the wind at some fraction, s, of the wind speed. Of course, the windsurfer may point his board in any direction he pleases and trim his sail to provide maximum speed in this direction, but whatever he does, his board will always have an additional component of velocity, $s\underline{w}$, in the wind direction.

When I apply the chapter 4 equation of motion to windsurfing, I must remember that the effective wind velocity experienced by the windsurfer is not $\underline{w}$ but is instead $(1\text{-}s)\underline{w}$ because of the onshore drift. This will change the apparent wind[2] and will result in a different equilibrium velocity over the water (compared with the equilibrium velocity we found for yachts). To see how fast sailboards can go, however, we are interested in determining the equilibrium velocity relative to the land, not the water, and so must add the drift velocity $s\underline{w}$ to the calculated equilibrium velocity. Before, we made no distinction between velocity relative to the water and velocity relative to land because we assumed that there was no net water movement. Here, we need to account for the fact that our sailboard is on water which is moving toward the shore. Einstein the sailor would approve of all this relativity. Math details are provided in note 3 for those of you who are interested in derivations; for those of you who are not the results will soon be presented conveniently in a graph.[3]

Before showing you these theoretical results for sailboard equilibrium speed, I need to take into account the fact that sailboards plane. This is easy: the hydrodynamic drag is much reduced, and so I will make the simplifying approximation that it is negligible for the purposes of determining equilibrium speed. Given this simplification, which I made in chapter 4 for iceboats, we start with the equation for iceboat equilibrium speed, already determined as equation (4.4) of chapter 4:

$$v_{eq} = w\left[\frac{c_L}{c_D}\sin(a_{vw}) + \cos(a_{vw})\right] \tag{7.1}$$

We do not accept this equation as it stands: we must include the consequences of water movement and leeway, discussed above, before it applies to windsurfing. In addition we must use lift and drag coefficients $c_{L,D}$ that pertain to sailboards. With a well-trimmed sail it has been found that windsurfers' sails can provide an L/D ratio of 6:1 or 7:1, so I adjust the yacht lift and drag coefficients (fig. 4.4) accordingly and obtain figure 7.10. Next, I substitute these coefficients into equation (7.1) and do the "relativity" thing: include board drift and determine the board speed relative to water and then relative to land. The results are shown in figure 7.11 for three different choices of the drift factor *s*.

How do I know what this factor should be? Well, I don't, and it will vary with windsurfing circumstances and with sailboard dimensions. So

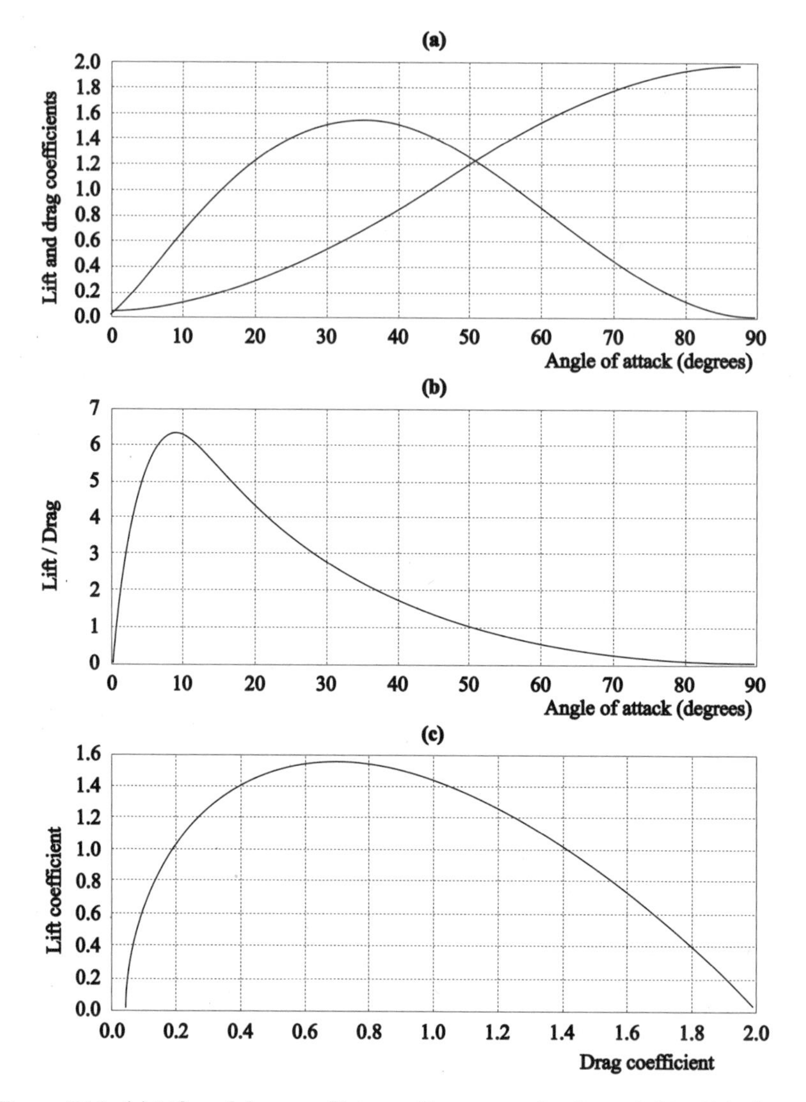

Figure 7.10. (a) Lift and drag coefficients $C_{L,CD}$ vs. angle of attack (AoA) in degrees. These are the coefficient values that I adopt for my sailboard calculations. (b) A comparison with the lift and drag coefficients for yachts given in fig. 4.4 shows the increased peak lift/drag ratio. (c) Sailboard lift coefficient vs. drag coefficient.

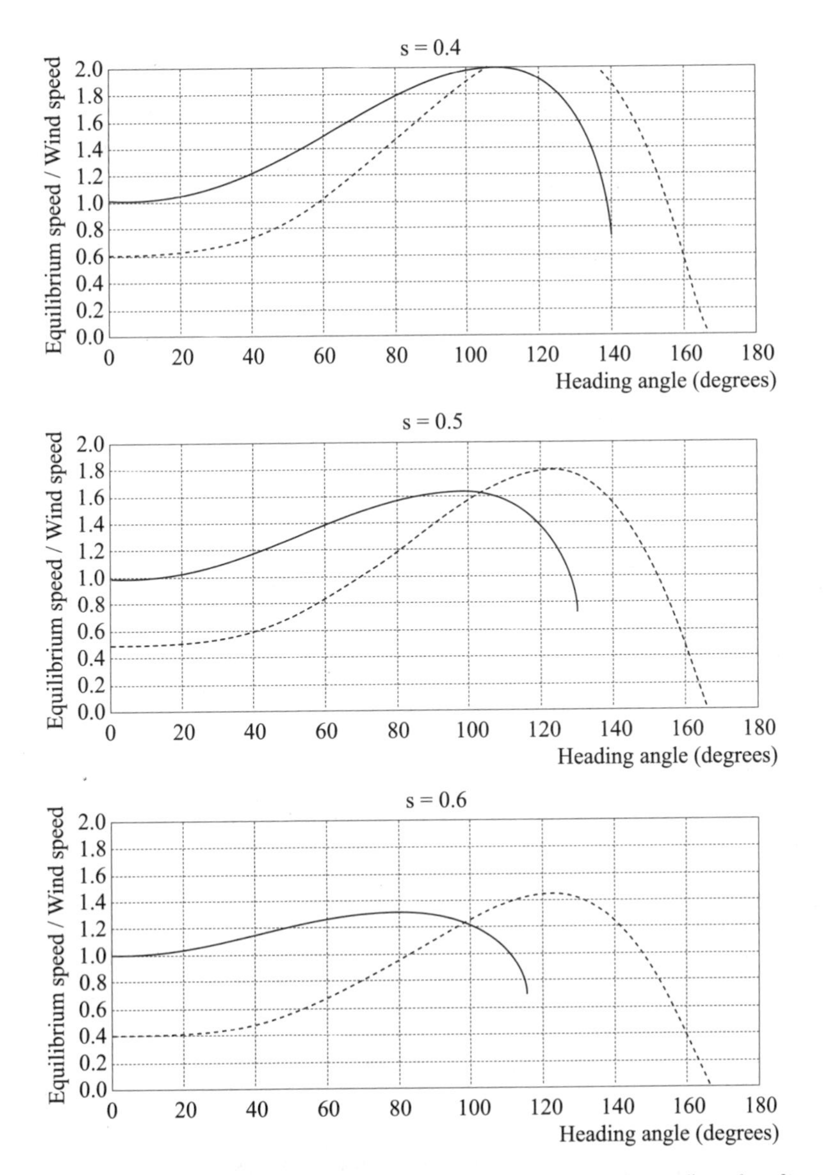

Figure 7.11. Windsurfer's equilibrium speed (divided by wind speed) vs. heading direction, for different "drift factors" (denoted by *s*). Dashed lines show the equilibrium speed relative to the water. Bold lines show equilibrium speed relative to land, which is what matters. The simple mathematical model shows that windsurfers can move faster than the wind, when on a beam reach or a broad reach. (A heading angle of 0° corresponds to downwind.) For each case plotted there is a maximum heading angle beyond which the board is effectively going backwards—in other words, the windsurfer is facing into the wind but drift is pushing him back. I assume that the windsurfer doesn't want to go there, so I won't either; I have not plotted this part of the curves.

I have chosen three middling values and plotted the results. If you disagree with my choice, you can substitute different s values in the equation (see endnote 3) and plot your own graph. The point is that it doesn't really matter because the plots show the same broad result; because my simple model is aimed at providing you with understandable physics, it will accept approximate answers. Figure 7.11 shows that windsurfers can go faster than the wind when heading on a broad or beam reach—in other words, when going across the wind or slightly downwind. A beam reach is convenient for those windsurfers (not a few, I suspect) who want to go as fast as possible while riding the front of a wave. Without accounting for board onshore drift, the calculations show that windsurfers' top speed is attained on a close reach—heading slightly to windward, as is shown in figure 7.11. This is not found to be the case in practice.

To summarize, we have seen here how the simple type of analysis used to gain insight into fore-and-aft yacht motion can be adapted and applied to sailboards. The adaptation is required to account for the onshore drift that we expect sailboards to encounter because they are unable to resist leeway and because they often ride waves and so drift with the water. Our results show that sailboards can go significantly faster than the wind that powers them. This observation suggests that sailboards might be able to "create their own wind" when on a beam reach, which in fact seems to be the case, as further calculations show.[4]

Pumping and the Katzmayr Effect

Pumping is the action of fanning the sail back and forth in such a way as to increase forward momentum. You may have observed windsurfers energetically wafting their sails and wondered how this action generates thrust (as an aeronautical engineer would call it, though some sailors call it "negative drag") for their rigs. A simple explanation follows. Pumping represents another difference between yachts and sailboards, at least in races, because for yachts pumping is controversially illegal whereas for sailboards it is *de rigueur*, even for Olympic competition. Careful legal definitions have been drafted*—after all, when does trimming a sail be-

*This is almost a pun, in the context of pumping, but I won't insist on it.

come pumping? A sailor is permitted to sheet in his sail and then let it out, but repeating the action constitutes pumping. Legal or not, the action is feasible only for small yachts because it requires a lot of energy. There have been many studies of the power expended by windsurfers when they pump their (relatively small) sails; the conclusion is that the rate of energy expenditure rivals that of a Tour de France cyclist or a distance runner. Pumping is important for windsurfers because it can get their boards planing very quickly.

For a sail that is being pumped, the action serves to oscillate the apparent wind angle of attack. The same action can be attained in a wind tunnel by fixing an airfoil and subjecting it to winds that vary in direction. The first person to perform wind tunnel experiments which showed that oscillating wind direction generated airfoil thrust was the Austrian aerodynamicist Katzmayr in 1922, and the phenomenon is now named after him.* Whether we regard the airfoil/sail as flapping in a constant wind, or the airfoil/sail as being subjected to an oscillating wind, is unimportant to physicists: in both cases the airfoil is subject to the same airflow. It is another example of relativity because it depends on where you are when you look at the action. A bug carried by the wind toward the airfoil sees a flapping airfoil approaching it. Another bug stuck to the airfoil sees wind coming first from this direction and then that, and then back again.

In the immediate aftermath of Katzmayr's demonstration, a number of aerodynamic engineers in several countries became very interested in the phenomenon and sought to reproduce his experiments and to provide a theory that explained the results. Their interest is understandable, given the potential consequences for aeronautics. Even today the Katzmayr effect is being studied in the context of micro-air vehicles (MAVs), tiny airborne sensor systems that appeal to the military and that are only now becoming technologically feasible.

The Katzmayr effect is observed in nature. Academic papers have been published with experimental results showing how dead trout can catch up with a boat—I kid you not. The flexible bodies of the deceased fish resonate in the vortex wake of the boat, gaining energy from the

*The U.S. Navy has an excellent website (see bibliography) discussing the Katzmayr effect in a technical but nonmathematical way, and with some excellent computer animations that greatly assist visualizing the sometimes complex airflow patterns.

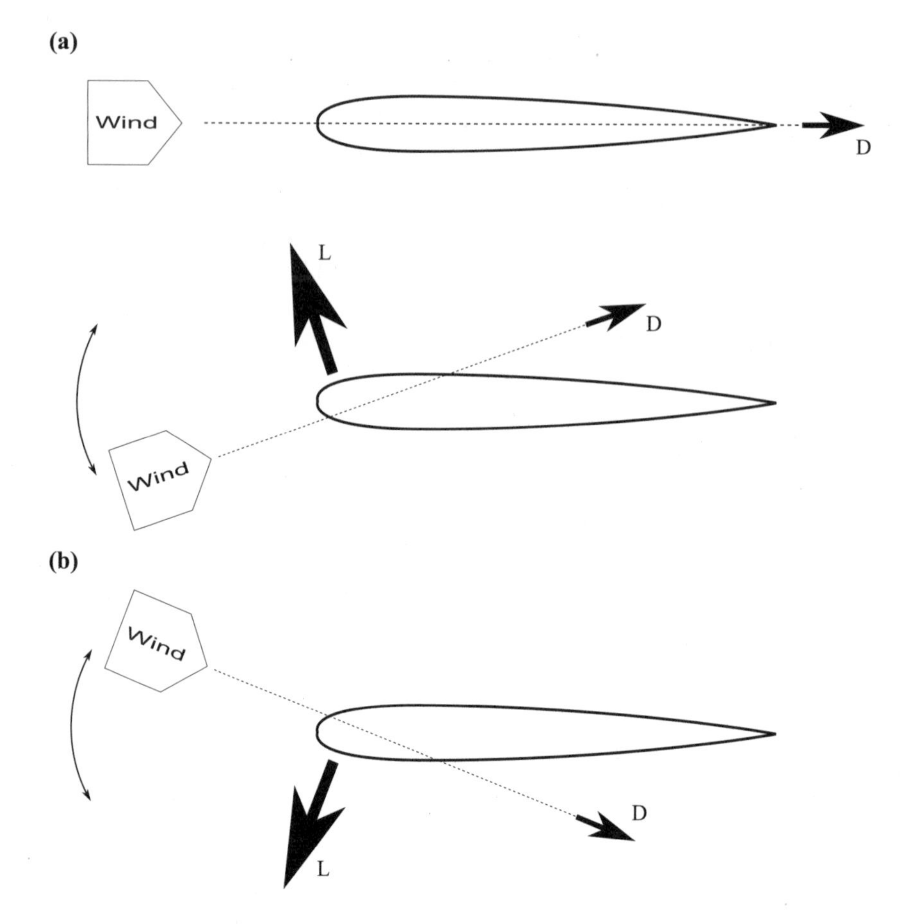

Figure 7.12. The Katzmayr effect. (a) A thin symmetric airfoil generates no lift when facing directly into the airflow. (b) If the angle of attack, or wind direction, changes, lift is generated. For an oscillating airflow direction the vertical component of lift adds to zero, over one cycle—i.e., no net lift. But the horizontal component is not zero; forward thrust has been generated.

vortices and thrust from the oscillating flow.* It is known that certain cetaceans (whales and dolphins) can exploit upstream oscillatory flow conditions to aid propulsion. The best-known natural example is provided by birds' wings, where lift, as well as thrust, is generated.

*A vortex is a rotating whirlpool of fluid—here, water—that detaches itself from a body (boat or airplane) moving through the fluid and is left in its wake. Vortices carry away energy from the moving body (they are a significant component of drag) and can persist for quite a while after the body has passed by.

Figure 7.13. Some windsurfers do not need waves to get airborne. Thanks to Stefania Bocheva for providing this photo.

The first explanation of how a bird wing can generate thrust came from two Germans, Knoller and Betz, in the early years of the twentieth century. Their explanation will be presented here because it is simple in concept and captures much of the physics without math. It is not the whole story, as we will see, but it works for me. Knoller and Betz adopt the viewpoint of the bug on the airfoil: fixed airfoil and oscillating wind direction, as shown in figure 7.12. Recall that aerodynamic drag acts in the same direction as the wind, while lift is perpendicular to the wind direction. For thin airfoils, it turns out that lift is very much greater than drag, and so, as a first approximation, we can ignore drag and say that the force on the airfoil is perpendicular to the airflow, as shown (fig. 7.12b). When the airflow direction is altered, the angle of attack is changed and lift is generated. You can see from the figure that this lift force has a forward component, which we call thrust. Half an oscillation cycle later, the wind direction is on the other side of the airfoil and generates lift in a different direction: the vertical components add up to zero but the horizontal (thrust) forces are always greater than zero. Hence, thrust results from an oscillating airflow.

Figure 7.14. Is this windsurfing or hitting a mine? I am grateful to Rich Swanner for these photos.

To simplify the explanation, the airfoil of figure 7.12 has been made symmetric. Real airfoils are cambered so that they can generate lift even at zero angle of attack, and this asymmetry can result in net lift as well as net thrust. Birds' wings are still more complicated because they adjust their pitch, so birds are fine-tuning the oscillating angle of attack throughout the wing-beat cycle. Windsurfers' sails are simpler, and yet pumping them to obtain thrust is still an acquired technique that involves shifting weight distribution along the board as well as brute-force manhandling of the sail. It requires finesse as well as strength and balance, and it can be very effective.

The main simplification of the Knoller and Betz explanation is that it does not discuss the role played by vortices. I will say something about vortices in the appendix, that correctional institution for wayward or overly mathematical theories of lift, but no more about them here. Except this: whenever an airfoil changes angle of attack, it generates vortices. These take energy away from the wing: recall all the power that a windsurfer must exert when pumping the sail. The energy of these vortices can be, in part, picked up by an airfoil that is further downstream (recall the dead trout). Some proposed MAV designs involve two wings, one of which is behind the other to mop up some of the energy that is shed by the front wing and generate thrust in the process. I wonder if a windsurfer can slipstream a competitor in a race (as the competitor pumps his sail) and, dead-trout-like, gain on him without expending so much energy?

Whether she pumps sails or shifts her weight, a windsurfer is able to do things that a sailor can only dream of (figs. 7.13–7.14)—or dread.

First rate: *Excellent, unsurpassed.* From the sixteenth century the British Royal Navy rated warships according to the number of cannons they carried. First rate corresponded to the largest and most formidable ships of the line; fourth- to sixth-rate vessels were normally considered too weak to be in the line.

Knock into a cocked hat: *To beat the opposition convincingly.* Originally a "cocked hat" was the small triangular space resulting from the intersection of three lines on a chart drawn when a ship's position was determined by three bearings.

Pipe down: *Be quiet.* The last call of the bosun's pipe, at the end of the day, signaling lights out and silence.

Toe the line: *To comply with what is expected.* When called to attention, a ship's crew would line up with toes touching a seam in the deck planking.

Appendix: Lifting the Veil

Mother Nature and Mathematics

The American Nobel prize–winning physicist Richard Feynman, a great lecturer and popularizer of science, once said that if we need math in order to explain a complex idea in science, then we don't really understand the idea. Feynman was a master of explanation, both technically to physics students and nontechnically to the general public, but he met his match when it came to explaining the concept of spin. By *spin* I do not mean the current phrase for political adumbration—saying a lot about nothing in a way that defies any clear and rational explanation. Instead, I refer to the property of elementary particles such as electrons and protons. These particles and others possess a "rest frame angular momentum" that physicists have given the name of spin, by analogy to a spinning top. This analogy is known to be wrong and misleading, but still physicists use the term *spin,* and physics students still emerge from their lectures thinking about electrons and protons as little spinning tops. Feynman sought to cut away the misconceptions with his usual pristine prose but was reduced to using math. He duly concluded that he (and by implication the rest of the physics community) didn't understand spin properly.

Bear with me: all this stuff about subatomic physics is relevant to an appendix on the theory of aerodynamic lift, if only by analogy. As with spin, aerodynamic lift is understood by physicists mathematically, and so far, Mother Nature has revealed her secrets to us on both these subjects using *only* the language of mathematics. Unfortunately, most people don't speak math, and so, for us physicists to explain either subject in a nontechnical manner, we must resort to simplifications and analogies. Just as the concept of spinning tops is misleading as an explanation of subatomic spin, the explanation of aerodynamic lift in terms of appealingly simple analogies is also misleading.

With lift, however, there is more than one misleading analogy. For this reason, most experts start off by debunking the analogies (see, for example, the NASA website on the subject of aerodynamic lift, listed in the bibliography). The multiplication of different analogies and simplifications (momentum flux, Bernoulli, Coanda, Venturi, . . .) plus all the debunking has led to an understandable bafflement among the general public about what underlies the difficult physical concept of aerodynamic lift. It is not my intention here to provide a detailed mathematical explanation—that would require a separate book, and most of you would rather undergo root canal work than read it. Instead, I will continue the philosophy behind this book by attempting to get across the physical principles that give rise to lift. Full appreciation of this appendix requires a greater knowledge of physics than was necessary for the main text, but I will make every effort to explain the concepts in a style that is accessible to nonexperts. Math such as I require is again relegated to endnotes. I will point the way to expert technical accounts for those who want to really get their teeth into the subject, but here, as in the main text, the emphasis will be on physical explanation, not mathematical calculation.

Names Picked Up and Dropped

Those inquisitive members of the sailing fraternity who are keen to learn about the physics of sailing have been subjected to a barrage of names, picked up over the years from a number of (often conflicting) accounts of how sails generate lift from the wind. The problem is correctly perceived as being closely related to the simpler problem of lift in airplanes. The names that you might have encountered in your investigations include Newton, Bernoulli, Euler, Navier, Stokes, Coanda, Magnus, and Flettner.* Perhaps surprisingly, given the dominance of Germans in explaining to the world of physics how lift occurs, only the last two of my list are Germans, and their involvement in the physics of sailing is only peripheral.† The Frenchman Navier and the Englishman Stokes together gave us the set of equations that describe with great generality the physics of fluid flow. It was the Navier-Stokes equations that I had in mind earlier when

*Rankine, Froud, Betz, von Karman, and many others contributed to our present understanding of fluid dynamics and lift, but the names in the text are the most significant for our purposes.

†Actually the full, gory, detailed understanding of aerodynamic lift, in terms of vortices, was largely a German achievement. Perhaps the complexity of the physics (as well as the math) has kept their names out of the popular literature. For the record, Joukowsky (or Zhukowsky), Kutta, Betz, Prandtl, Munk, and Lanchester figure prominently—a Russian, four Germans, and an Englishman.

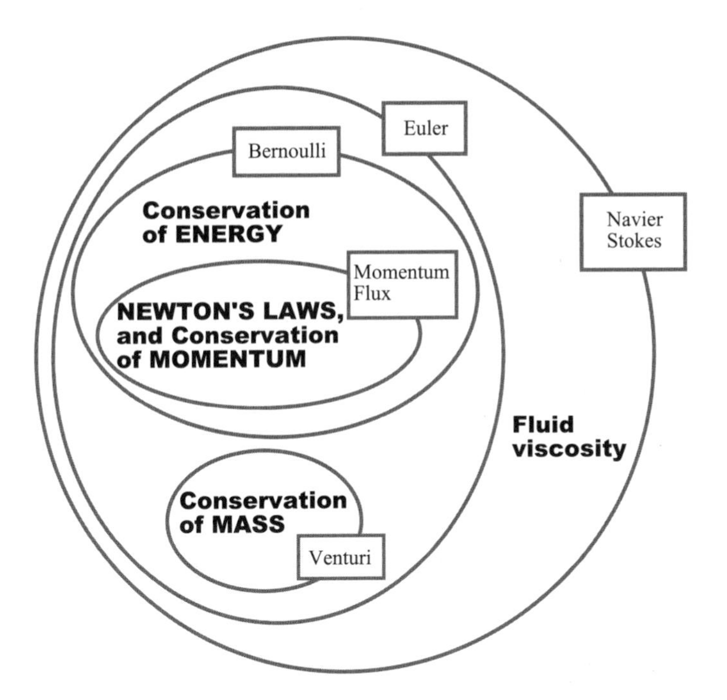

Figure A.1. Four physical principles and five sets of equations used to describe fluid dynamics. Lift and drag forces can be calculated very accurately from the Euler equations, and approximately from the Bernoulli and momentum flux approaches.

referring to root canal work. These equations are so horrible to work with* that a whole field of physics, computational fluid dynamics, has arisen to solve them, and most physicists prefer one or another simplifying approximation. This simplification process has led to misapplication and confusion, here to be expurgated.

The various approaches to explaining fluid dynamic lift force, and the physical principles that they embrace, are shown schematically in figure A.1. The more principles that an approach includes, the better it approximates reality. However, the history of explanations for lift force is littered with misapplications of good equations and principles, resulting in confusion.

Navier-Stokes equations apply wherever there are fluids: they take into account energy conservation, momentum conservation, mass conservation, and fluid viscosity. When restricting ourselves to aerodynamics we can neglect the effects of viscosity, except for the very thin boundary layer of air that surrounds the airfoil. With this approximation, we move from Navier-Stokes to

*Unless coupled partial differential equations light up your life.

Euler. The Euler equations* are "cod-liver-oil" difficult, rather than the more severe "root-canal" difficult—you would rather drink a pint of cod-liver oil than deal with them. The solutions to Euler's equations provide excellent agreement with wind-tunnel experiments on airfoil lift and drag, thus giving physicists confidence that they have captured the physical principles of aerodynamics in detail. So far, so good. However, because Euler's equations are difficult (because fluid dynamics is difficult), it lacks an intuitive appeal, especially when explaining aerodynamics to nonexperts—meaning other physicists as well as the general public. Consequently, a number of attempts have been made over the last hundred years or so to provide more intuitive, if more approximate, equations that capture enough of the underlying aerodynamics to be plausible. The trouble is that none of these attempts has been very successful, until recently.

There are two main factions in the often heated debates surrounding aerodynamic lift, which I will discuss in turn: the "Bernoulli faction" is populated by those who think of lift force in terms of pressure, whereas the "momentum flux faction" consists of those who prefer a more direct application of Newton's laws. The approach of the Bernoulli faction takes several forms and appeals to Bernoulli's equation.† At their best, the Bernoulli faction is right: its adherents apply Bernoulli's equation—and one of Euler's—to airfoils, and from the resulting streamlines they can calculate with some accuracy the lift force that is appropriate to that airfoil. The correctly applied Bernoulli equation also provides a little physical insight, though only a little, which is not forthcoming from a study of Euler's equations. (People tend naturally to think in terms of mechanics, and forces, because these concepts are apparent in our everyday lives. Unfortunately, fluid mechanics appears disconnected from our intuitive understanding of mechanical apparatus and forces, and so we try to bend the math of fluid mechanics to yield a mechanical model within our comfort zone.) Bernoulli's equation can be derived from energy conservation but also from an application of Newton's laws:

$$p_s + \frac{1}{2}\rho v^2 = p \qquad \text{(A.1)}$$

Here, p_s is the static pressure of the fluid (for example, atmospheric pressure), and p is the total pressure, which is found by Bernoulli to be constant. The second term on the left is dynamic pressure: ρ is fluid density and v is the velocity of a particle of fluid. This equation (which applies *only along a streamline*—

*From the fertile mind of the famous and brilliant eighteenth-century Swiss mathematician Leonard Euler (pronounced *Oiler*).

†Daniel Bernoulli, a Dutch/Swiss mathematician, published his seminal work on fluid dynamics in the 1730s.

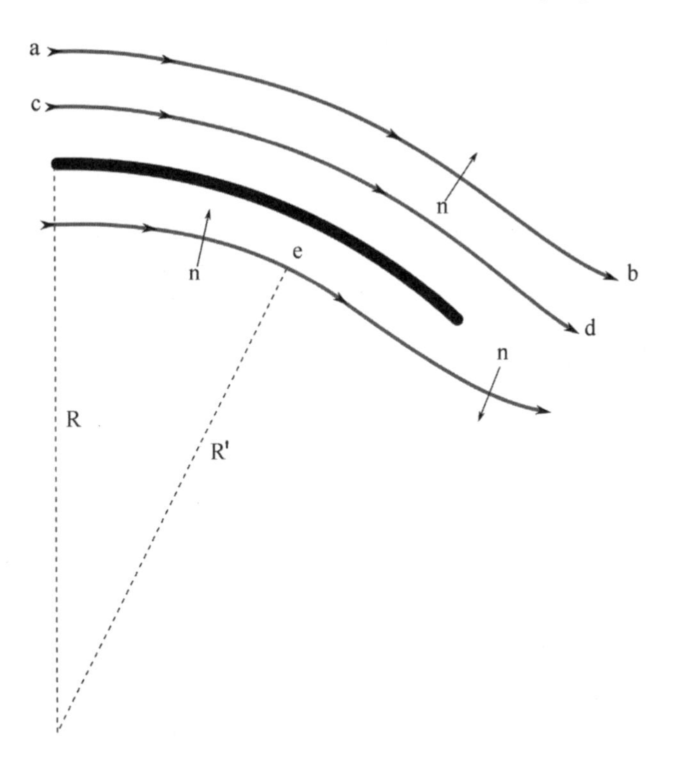

Figure A.2. Bernoulli's equation (A.1) relates fluid speed and pressure at different points on an individual streamline—for example, points *a* and *b*, and points *c* and *d*, but not points *a* and *d*. Equation (A.2) relates different streamlines. The dashed line *R* is the local radius of the airfoil (bold); the lowest streamline follows this airfoil shape and so has the same center of curvature, but a smaller radius, R', at point *e*. The normal (i.e., perpendicular) direction of increasing pressure is shown at several points by short arrows.

a fact that is too often forgotten) tells us that when the velocity of a section of the fluid is high, then the static pressure is low. When the velocity is low, the pressure is high. These statements are true, but are very often misapplied when explaining lift, as we will see.

The Euler equation of relevance to us here is

$$\frac{dp}{dn} = \rho \frac{v^2}{R} \qquad \text{(A.2)}$$

Here, n is the direction that is normal to the tangent of a curved streamline, as shown by arrows in figure A.2, and R is the local radius of curvature of that streamline. This second equation applies across streamlines and tells us how the fluid pressure varies between streamlines. Note that regions of high curvature (small R) have larger pressure gradients and hence exert larger forces. Thus, the pressure gradient below the airfoil of figure A.2 is greater than that above because the radius of curvature below is less. The gradients immediately above and below the airfoil are both positive (n points outward), and so we deduce that air pressure is greater at the lower surface than at the upper surface of the airfoil. How so? Because the pressure far below and way above the airfoil must be the same static atmospheric value, p_0 (neglecting the effects of gravity on the atmosphere). Imagine that you are following a straight line up from the ground, through the airfoil surface, and beyond to the stars. The positive pressure gradient that you feel means that pressure rises from p_0 to some higher value p_+ as we approach the lower surface, and rises from some lower value p_- to p_0 as we recede from the upper surface. Greater pressure below than above the airfoil means lift—*voila!*

This is how the "pressure people" (the Bernoulli faction) explain lift. The argument is a little contorted but sound. Attempts to simplify the reasoning have led people astray, as we will now see.

Path Difference

A very common and wrong appeal to Bernoulli, when explaining lift, goes like this: "*The air travels farther over the convex, upper side of an airfoil than it does over the concave side, and so the air speed is greater above the airfoil than below. From equation (A.1) we know that higher speed means lower pressure, and vice versa. So the pressure above an airfoil is reduced, and the pressure below is increased. Multiplying pressure by area gives vertical lift force.*" This explanation (illustrated in figure A.3) is incorrect for a number of reasons.

First, in saying that the air has "farther to go" there is an implicit assumption that the air above and below the airfoil must match up. That is, two particles of air that are separated by the front of the airfoil—one traveling over and one underneath—meet again at the trailing (leech) end. This is a misconception; wind-tunnel tests have shown that two such particles do not need to meet up again. It seems that this misconception arose early on, and from an eminent source: Ludwig Prandtl, no less. A diagram in his book shows clearly that he assumed the particles would meet up at the leech. Even if they did meet up, there is no reason to assume that the upper particle travels farther (and therefore faster) than the lower particle. True, some airfoil shapes (such as that of figure A.3) produce a larger traveling distance, but the difference is negligible for thin airfoils such as sails. And what about planes flying upside down? This explanation does not work for them.

The second error in the "path difference" argument is in applying Bernoulli's equation (A.1) to the problem. This equation is restricted to an individ-

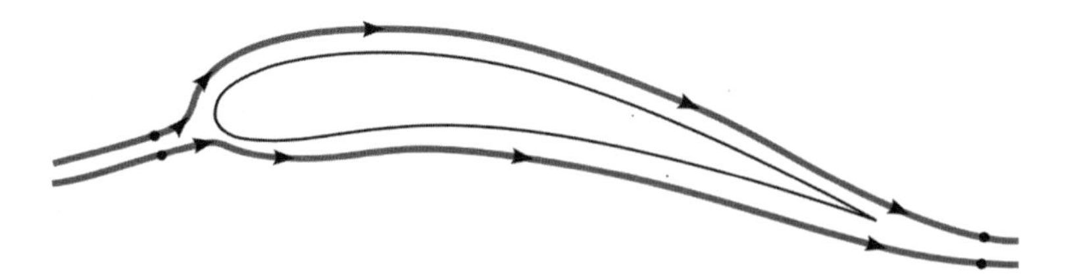

Figure A.3. The path difference "explanation." Two particles (dots at left) that become separated by an airfoil are supposed to align again once they have passed the airfoil as shown. Given the airfoil shape, this means that the upper particle moves faster. So, misapplying Bernoulli, it is concluded that pressure is reduced near the top surface of the airfoil, resulting in lift. In fact, it is not true that the particles realign downstream of the airfoil, as wind-tunnel tests have shown.

ual streamline, as noted above, and is not to be applied across streamlines—for that we need equation (A.2). The problem with equation (A.2) is that to obtain quantitative results the equation needs to be integrated, and it is not at all obvious how to do this integration, as we will see.

The third error is very misleading: the assumption that velocity gradients lead to pressure differences puts the cart before the horse. In fact, the pressure differences give rise to velocity gradients, and not the other way around.*

The correct application of the Bernoulli approach, we have seen, is to apply equation (A.2). To obtain an expression of lift force from this equation, we need to integrate the pressure derivative over area to obtain the total force acting on the airfoil. The vertical component of this force is lift. The first attempts to perform this integral led to an alarming result, now known as *d'Alembert's paradox:* the total force that results for an ideal fluid, if we neglect vortices, is zero, regardless of the shape of the airfoil. The problem here is not the assumption of an ideal fluid (one with no viscosity) but of neglecting vortices. It turns out that vortices are essential to a full and complete understanding of lift (I will have more to say about vortices below). But vortices are root-canal complicated to work with because they are shed from the airfoil and spill back into the wake. So what surface do we integrate over to obtain the total force?

A common and reasonable approximation turns out to be that of *circulation,* often found in the more technical articles which seek to explain lift force. The circulation concept, illustrated in figure A.4, amounts to assuming a bound vortex—one that does not shed from the airfoil. Including circulation yields lift force, to a reasonable approximation. The only issue I have with circulation is

*There are many members of the Bernoulli faction who reject the "path difference" argument and yet persist in saying that velocity differences lead to pressure differences (and hence lift).

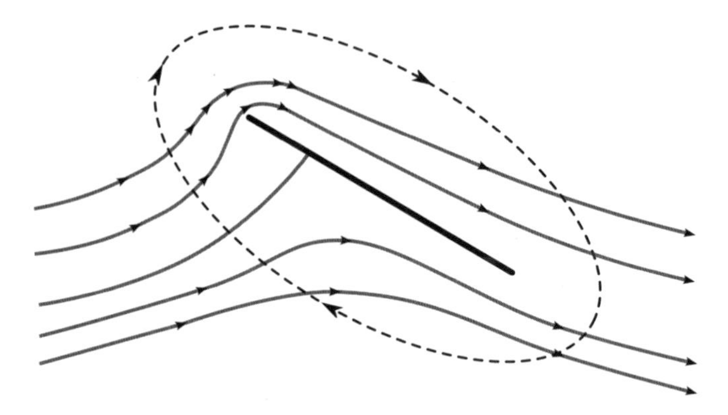

Figure A.4. Here, we have a straight airfoil inclined at an angle of attack to produce lift. Air particles that follow streamlines (gray) above the airfoil are observed experimentally to move faster than particles following streamlines below the airfoil. The streamline shown without arrows separates the streamlines over and under the airfoil. The velocity field (the distribution of air speed around the airfoil) can be modeled as a steady flow field—with constant air speed and horizontal everywhere—plus a circulation pattern, here clockwise and shown as a dashed line. In fact, there are many such circulation lines, with speeds that diminish with distance from the airfoil. The steady flow part of the velocity field contributes nothing to airfoil lift.

that it is often put forward as an explanation, rather than as a mathematical description, of the velocity distribution around an airfoil. It is a tool for calculation; it does not explain *why* the velocity distribution is the way that it is.

To summarize, certain members of the Bernoulli faction misapply in various ways Bernoulli's excellent equation. Other members provide us with a good approximation to the lift force of an airfoil but provide little explanation of why it works.

The Venturi Nozzle

Before leaving our discussion of the approach of the Bernoulli faction, we should look at the approach of an offshoot—the Venturi school*—who offer a simpler explanation of lift. The simplicity of this explanation is seductive, but again, the price to be paid is bad physics. There is certainly a Venturi nozzle effect, as shown in figure A.5; the problem is that it is misapplied when utilized to explain lift. Venturi nozzles are used to determine the airspeed of a plane by

*Named after Giovanni Battista Venturi, Italian physicist of the late eighteenth century.

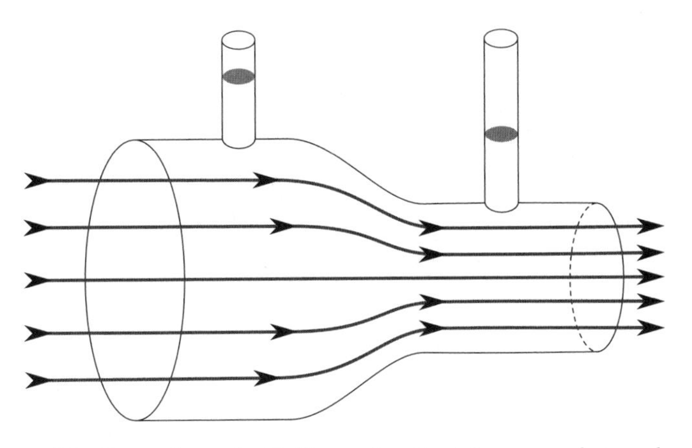

Figure A.5. The Venturi nozzle. Fluid entering the tube moves faster where the diameter is reduced. According to Bernoulli's equation, the pressure is reduced in this region.

passing air through a constricted tube. The mass of fluid is conserved, of course, and so its speed must vary inversely with the tube's cross-sectional area.[1] We know from equation (A.1) that pressure is lower where air speed is higher. The Venturi effect has been misapplied to explain lift force, as follows. "*Air moves faster over the top surface of an airfoil, which is designed to create this effect. We know from the Venturi tube that faster air results in lower pressure and so there is a lifting force produced.*" The trouble is that the air through which an airplane flies, or through which a boat sails, is not constricted to a tube. There is no outer boundary to the airflow. Furthermore, in the Venturi approach, as with the "path difference" approach, pressure reduction is seen to be a consequence of speed increase rather than the other way around. Also, this argument ignores what happens at the lower surface of an airfoil. It cannot, for example, explain how a flat plate that is inclined to the flow at a low angle of attack can generate lift.

Paper Games and the Coanda Effect

There is a well-known demonstration of the Bernoulli faction's viewpoint that lift is due to increased velocity in the streamlines. This demonstration is interesting for three reasons: (1) it has nothing to do with increased velocity in the streamlines; (2) a simple counterexample clearly shows that it has nothing to do with increased velocity in the streamlines; and (3) it in fact supports the view of lift by the opposing momentum flux faction.

Consider figure A.6. You hold a sheet of paper between your fingers hori-

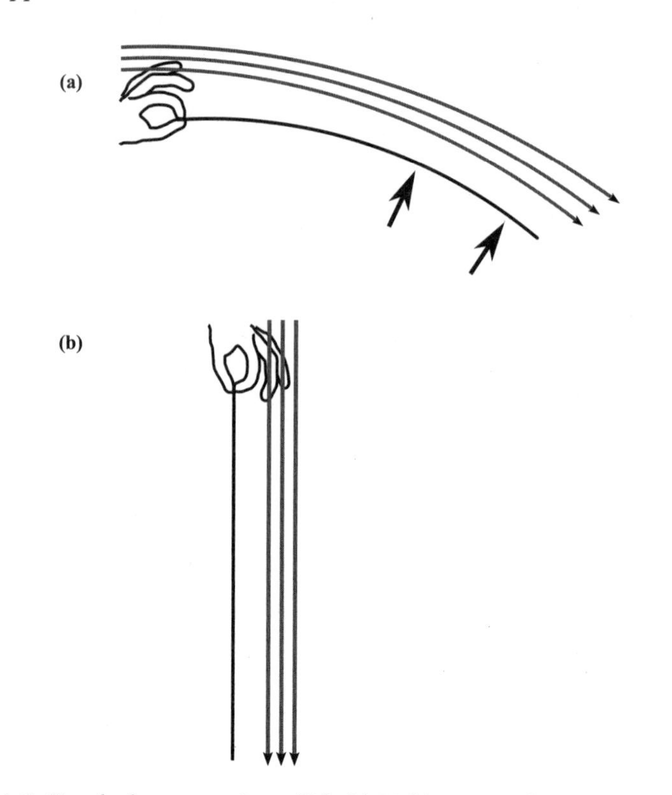

Figure A.6. Simple demonstration of lift. (a) Hold a piece of paper horizontally between your fingers so that it curves and blow over the top of it. The paper will lift up. (b) Hold the same piece of paper vertically and blow down one side. There is no movement. These two demonstrations taken together show that a speed difference on the two sides of an airfoil is not enough to generate lift: airfoil curvature is required.

zontally so that it curves and blow over the top of it. The paper is observed to rise. You are supposed to conclude: "*The increased airflow speed above the paper causes reduced pressure above the paper, owing to the Bernoulli equation (A.1), and so creates lift, which we observe when the paper rises.*" Baloney. Consider a second paper experiment in which the paper is held vertically and straight (also shown in figure A.6). Blow again along one side of the paper—no lift. Why? Because increased speed does not cause reduced pressure. One more time: it is the other way around—reduced pressure causes increased speed. I commend these simple experiments to you; they are very easy to perform and are quite clear in their outcome.

So what does cause the curved paper to rise? It certainly is subjected to a lift force, but the vertical paper experiment shows that the view of the Bernoulli

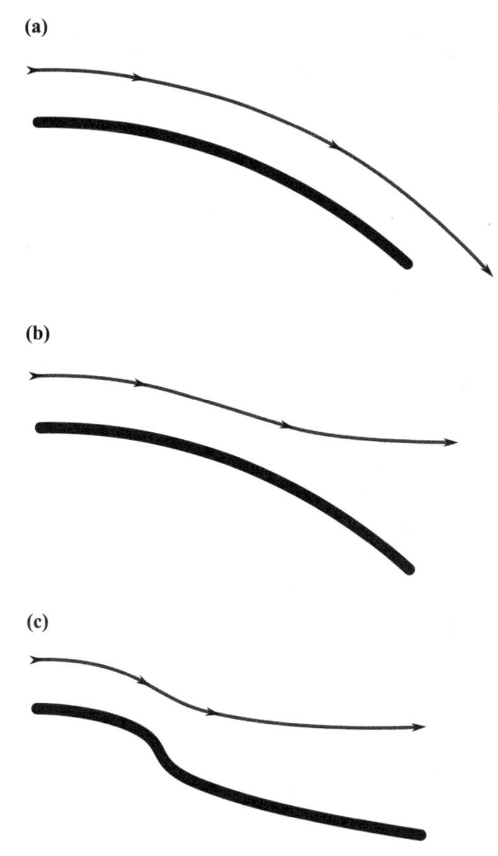

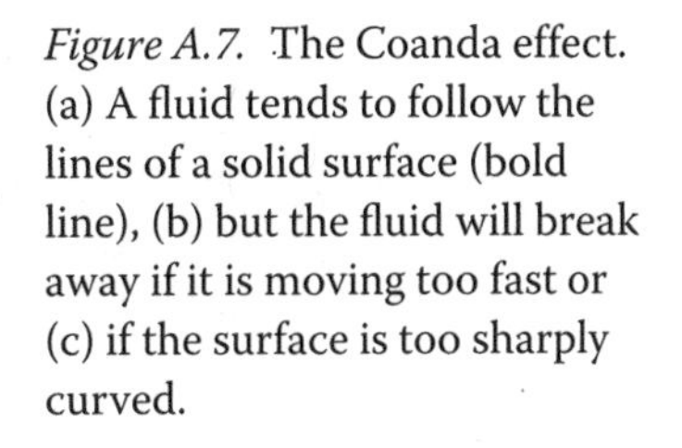

Figure A.7. The Coanda effect. (a) A fluid tends to follow the lines of a solid surface (bold line), (b) but the fluid will break away if it is moving too fast or (c) if the surface is too sharply curved.

faction is wrong.* There is a phenomenon known as the Coanda effect,† illustrated in figure A.7, which accounts for the airflow curvature. Fluid moving parallel to a solid surface tends to follow the surface so long as the fluid speed is not too great and the surface curvature is not too great. The behavior appears to be a skin friction effect and is readily demonstrated (fig. A.8).

The Coanda effect accounts for the curvature of the airflow that passes the curved paper, but why does this produce lift? And why not when the paper is straight? A simple and compelling explanation is provided by the modern version of momentum flux.

*To be fair, this is the view of a misinformed subset of those whom I describe as the Bernoulli faction, though it is a large subset judging by the number of times this experiment is promulgated in their support.
†Henri Marie Coanda, early-twentieth-century Romanian aerodynamicist. The effect named after him was known for at least a century prior to his investigations.

Figure A.8. A familiar example of the Coanda effect. Photo by the author.

Momentum Flux Redux

We used a naïve version of the momentum flux approach in chapter 2 to explain how square-riggers derive drive from the wind. The most naïve version, as pointed out in chapter 2, simply supposes that the air particles impinge on the lower surface of an airfoil and transfer momentum to it; the vertical component of this momentum is lift.* This approach is a very good approximation to the truth in regions of high speed and low air density, as when the space shuttle reenters the atmosphere, but for aircraft near the earth's surface and for yachts it is not good. Our version was a little less naïve because we recognized that air passing over the top surface of the airfoil is also deflected, though we could provide only a rough argument as to why. To account for *all* the deflected air, recall, we made use of the effective sail (airfoil) area $A = \pi h^2/4$, where h is the maximum dimension of the sail. Even so, the lift forces calculated are only in rough agreement with measured results (though good enough to show how

*Note how different this naïve theory is from the Venturi misapplication just discussed. In one case only the upper airfoil surface contributes to lift; in the other only the lower surface contributes. In fact, both surfaces matter.

square-riggers work). This method is unable to explain how some airfoils with zero angle of attack can generate lift.

Over the last few years a number of physicists have begun to advocate on behalf of a more sophisticated version of the momentum flux approach and have wrested the popular explanation of lift from the Bernoulli faction, who dominated for so long, leading to much confusion.* Yet the Bernoulli faction (specifically the "Euler contingent"—recall equation [A.2] and the reasoning behind figure A.2) is correct. What was needed was a more intuitive explanation of figure A.2. Consider this figure once again from a Newtonian point of view. The curved airfoil deflects air flow due to the Coanda effect: streamlines near the airfoil run parallel to it. This means that the nearby streamlines take on the curvature of the airfoil. Sir Isaac Newton, on seeing a particle of air moving along a curved trajectory, would conclude that the particle is accelerating. This centripetal acceleration is toward the local center of curvature and can be produced only by a pressure gradient,† so the air pressure must be increasing with increasing distance from the center of curvature. The rest of the argument is the same as before: imposing the boundary condition (that air pressure far away from the airfoil must be atmospheric pressure) we obtain higher pressure beneath, and lower pressure above, the airfoil. Hence, a lift force is produced as a consequence of streamline curvature.

Note that from the momentum flux viewpoint it is streamline curvature that matters, not airfoil curvature. The airfoil will influence streamline curvature, but lift can still be produced by a flat airfoil if the angle of attack is appropriate. Thus, in figure A.4 we saw a flat airfoil produce a disturbance in the airflow velocity field, resulting in a net concave-down curvature. This is in the right direction to produce lift from the centripetal acceleration argument. So now we know why airflow above as well as below an airfoil can produce lift. We are now able to reiterate the earlier naïve momentum flux statement with more authority: airflow is redirected downward by an airfoil (because of the

*Bernoulli explanations dominated the teaching of lift at the introductory level for the past 80 years. Interestingly, for 20 years previously (i.e., for the first two decades after the physics of lift was understood by experts in terms of vortices) teachers of aerodynamic lift preferred a momentum flux approach close to the modern one, as is clear from the writing of Otto Lilienthal, an early German aviation pioneer. For clear expositions of the modern approach using painless math, see the papers by Babinsky and by Weltner and Ingelman-Sundberg cited in the bibliography. For a more technical overview that shows how the Bernoulli and momentum flux approaches fit in with the expert Euler theory, see the paper by Auerbach.

†There is no other source of force because gravitational effects are negligible and we are neglecting molecular forces except within a very thin boundary layer near the airfoil surface.

Coanda effect and airfoil shape and orientation), and this downwash creates a reaction force (by Newton's Third Law), lifting the airfoil.

The curved-streamline view gets around a weakness of the naïve momentum flux approach: a cambered airfoil at zero angle of attack will create curved streamlines and hence generate lift. Recall that, in the naïve version, zero angle of attack meant zero lift. We can also see qualitatively how the more sophisticated momentum flux approach leads to an understanding of stall* and of aircraft that can fly upside down. In figure A.9 we see that an airfoil experiencing stall (at too high an angle of attack) causes streamlines that are less curved than the same airfoil with a lower angle of attack. Momentum flux says nothing about vortices, but we can understand how the streamlines detach from the upper surface of the stalling airfoil from the Coanda effect. The curvature required of the streamlines at the front of the airfoil is too much, and the air particles can no longer run parallel to the airfoil surface. Once detached, the streamline radius of curvature is greater than it would be without detachment. For a cambered airfoil flying upside down, we see that lift can result (albeit inefficiently) if the angle of attack is large enough. The curvature is wrong for the lower streamlines but is in the right direction for the upper streamlines, with a lower radius (increased pressure gradient, and hence lift) for higher angle of attack. For very large angles of attack, presumably, drag increases so much that forward motion is compromised and so flow speed is reduced, thus reducing lift.

The theory of sail lift is more complicated that that of airplane wing lift, but we see that the new momentum flux argument can still be applied and that we gain insight from it. Thus, for example, we can examine the slot effect of reduced mainsail drag† via both the Bernoulli and the momentum flux viewpoints,‡ as shown in figure A.10.

*At high angles of attack, the airflow cannot follow the airfoil surface and separates from it forward of the airfoil trailing edge. The switch from smooth flow to separation can be sudden, and results in rapid loss of lift, which we call "stalling."

†Slot-effect aerodynamics were first sorted out less than 40 years ago, by Arvel Gentry (see the bibliography).

‡There is a Venturi misapplication here, too. The slot between jib and mainsail looks, to some people, like a Venturi tube constriction, and so they claim increased airflow speed in the slot, and "therefore" reduced pressure and greater lift. However, wind-tunnel tests show that air flow speed in the slot is reduced, not increased.

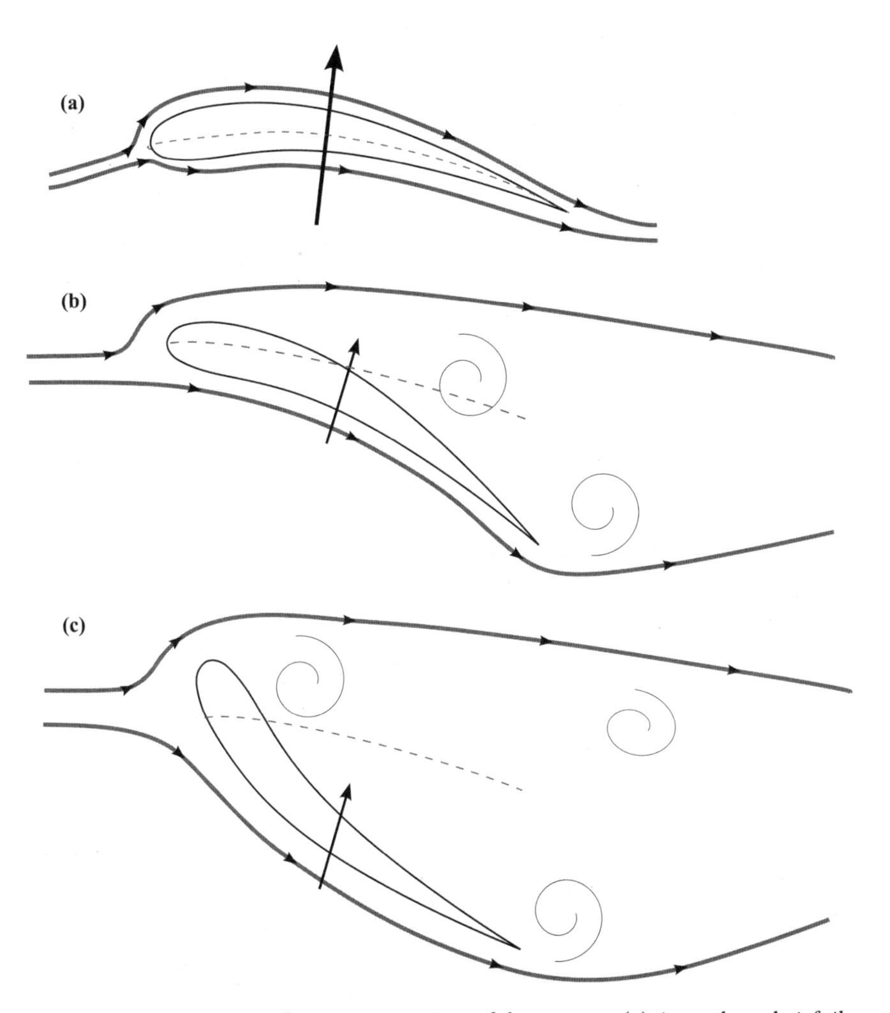

Figure A.9. Momentum flux interpretation of three cases. (a) A cambered airfoil with a small angle of attack. The net streamline curvature in the vicinity of the airfoil (suggested by the dashed line) is concave-down. Thus, air pressure exceeds atmospheric pressure under the airfoil and is lower than atmospheric above the airfoil, resulting in lift (arrow). (b) The same airfoil stalls at a steeper angle of attack. There is still a net concave-down curvature, but the radius of curvature is greater, and so the pressure gradient and lift force are reduced. (c) The same airfoil upside down. The net curvature provides a slight lift but requires a steeper angle of attack.

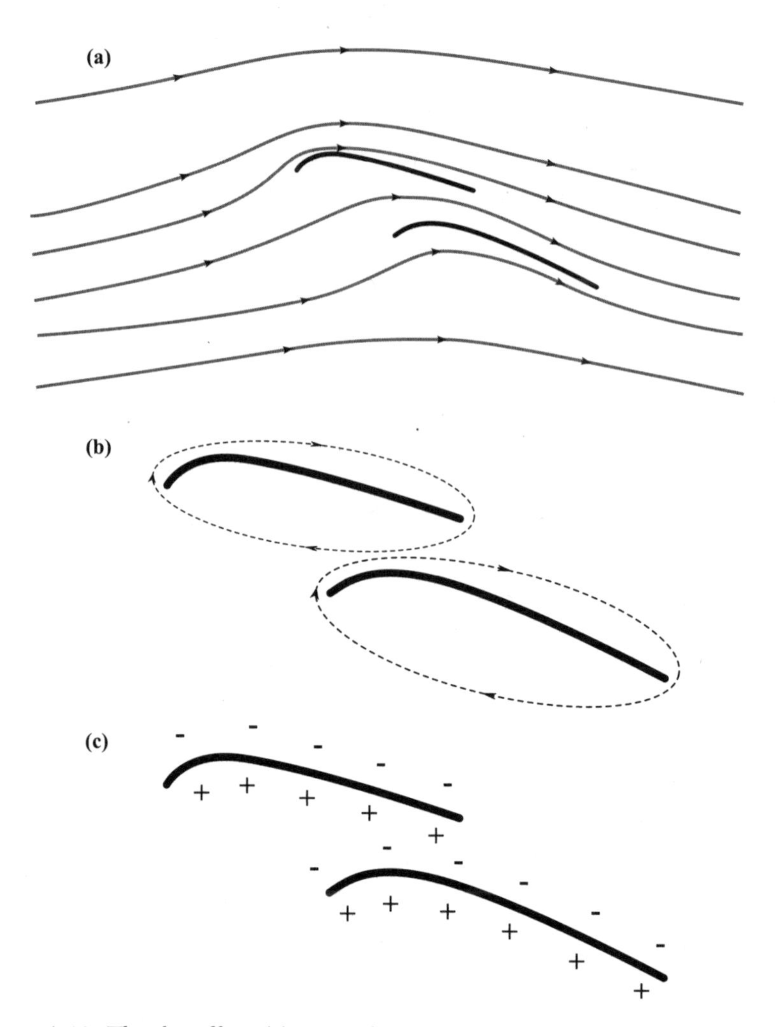

Figure A.10. The slot effect. (a) Streamlines. (b) The Bernoulli view: circulation fields around both sails cancel each other in the slot, so air speed is reduced. Lower speed helps to prevent separation in the lee of the mainsail, thereby reducing mainsail drag. (c) The momentum flux view: streamline curvature results in increased pressure at the windward surfaces of both sails and reduced pressure on their leeward surfaces. In the slot, these cancel. Air in the mainsail lee will not separate as easily as it would without the jib. Hence, less mainsail drag. Expressed more succinctly, jib downwash reduces the effective mainsail angle of attack, reducing mainsail drag.

One True Theory

You may have noticed that there is not a great deal of difference between the best arguments of the momentum flux proponents and those of the Bernoulli group. Perhaps this convergence of thinking is not too surprising because there is only one true explanation and both camps (if they are to produce the right answer) must approach it, albeit from different directions. One advantage of the momentum flux viewpoint is that it makes the underlying physics clearer for nonexperts. Thus, airfoil shape or orientation to the airflow influences streamline curvature (due to Coanda effect), creating pressure gradients perpendicular to the airflow (due to centripetal acceleration). From pressure gradients we obtain lift. The Bernoulli-Euler faction obtain the same pressure gradients but explain them as a consequence of circulation. It is not obvious from their explanations where circulation comes from, though few doubt that it produces the right answer.

Calculations of the lift produced by circulation show that it is divided into two parts. Recall that it is necessary, when calculating lift, to integrate the pressure derivative of equation (A.2) over the surface defined by the circulation field. The integral receives contributions from the fluid pressure at the surface and from the momentum flux of fluid moving through the surface. But how big is the surface? That depends on how much of the airflow velocity field is disturbed by our airfoil.* We know from experiments that the influence of an airfoil can extend far away, in all directions, and so it is sensible to take as large a surface as possible. But what shape should it be? It turns out that the shape is important. If we pick a cylindrical shape (so that the surface looks like a circle on the airflow diagram of, for example, fig. A.4), then the pressure component and the momentum flux component each contribute half the total lift. But if we change the shape of the surface we find that the relative contributions change (though the total contribution is unaltered by changing the surface shape).

This odd result suggests to me that the division between pressure and momentum flux is somewhat artificial, but it cannot be entirely so because in some cases the surface shape is chosen for us by the geometry of the situation we are investigating. For example, in a wind tunnel we know that the boundary is limited by the wind-tunnel walls. We expect that momentum flux would be more significant in this situation, and this is what we find from the circulation integral. For an aircraft flying over the surface of the earth, we expect at least some contribution from momentum flux; and indeed, Betz as early as 1912

*Recall (fig. 2.4b) that the momentum flux approach had just this problem, of how much air to include, in the form of the sail/airfoil effective area (which is not the same as the geometrical area).

calculated the force produced on the earth's surface by a passing airplane. On the other hand, there are situations where pressure contributes everything. Consider the central streamline in the Venturi tube (fig. A.5). There is no curvature here and so, according to the momentum flux approach, no pressure change. Yet there is indeed a reduced pressure in the constricted part of the tube. For determining aerodynamic (and hydrodynamic) lift forces, however, it seems that the idea of momentum flux is the best one available for conveying the complex physics of fluid dynamics in an intuitive way.

Enter the Vortex

At this point I have to yield to Feynman's dictum. I refuse to inflict upon you the mathematics of vortices—they are an order of magnitude more complex than the mathematics used so far in this book—and so I resort to a more heuristic explanation, and accept that it will not provide you with a full understanding. The consolation is that it is quite possible to provide a good intuitive feel for how vortices arise and act by using a few well-chosen diagrams. This is because the problem is essentially a geometrical one, and diagrams are good at explaining geometry.

So far I have only hinted at the three-dimensional nature of fluid dynamics. Most of the streamline diagrams that you have seen in earlier chapters are two-dimensional. It is as if the airfoil they represent is infinite in extent, so that we do not have to worry about what happens at the wing tip. But the wing tip is where, in a basic sense, all the action is. Consider the 2-D airfoil diagrams that you have seen to this point—for example, figure 2.4 or figure A.4. In a steady flow, of an airfoil moving through the atmosphere at uniform speed, these diagrams would remain stationary in time. The streamlines are unvarying, and nothing in the diagrams changes as time ticks by. Look at another section further along the wing and you see the same diagram. This cozy situation may apply in two dimensions, in the benign case of nonturbulent flow, but it does not apply in three dimensions. Consider what happens for an airplane wing that is of finite length (and, in my experience, most airplane wings do not extend to infinity). The three diagrams of figure A.11 should convey the "new" physics that arise in 3-D. I begin with airplanes and will graduate to boat sails later.

First, you can see that the air below an airplane's wing reaches around the tip to the top surface of the wing. It does this because the air beneath is at high pressure and the air above is at low pressure; the moving air is simply trying to equalize the pressure. It cannot move around the leading or trailing edges of the wing because the airflow impedes progress along these routes and so must move around the tips. Thus, air flows outwards along the lower surface of a wing and inwards along the upper surface. Second, you can imagine how the

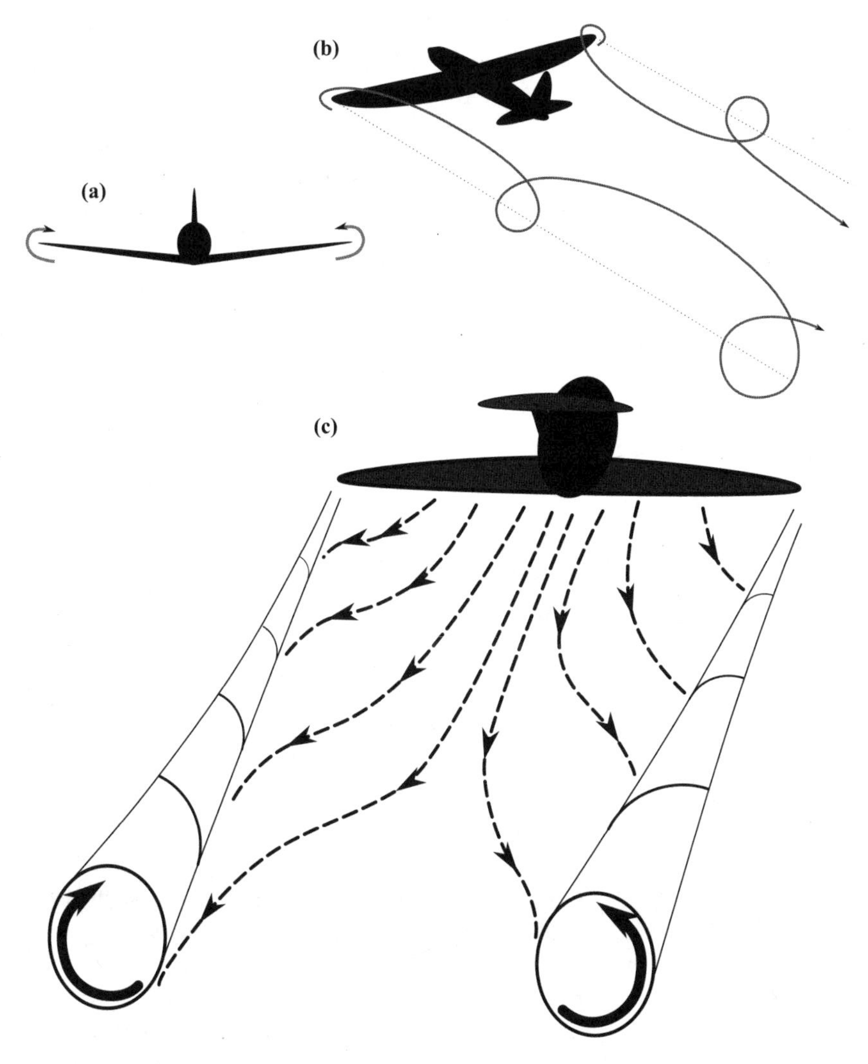

Figure A.11. (a) Air under the wing tries to move to the upper surface of the wing—from high pressure to low pressure. It can do so only by moving around the wing tips. (b) But the plane is moving forward, and so the rotating air is shed from the wing tips and spills backwards as vortices. (c) Air from the other parts of the wing are sucked into these vortices, which rotate in opposite senses, as shown. Such wing-tip vortices can persist for some time and for some distance in the wake of a large aircraft.

airflow causes the rotating air at the wing tips to become detached from the wing and spill backwards in swirling spirals—these are the wing-tip vortices. The left vortex (viewed from behind the airplane) rotates clockwise and the right vortex rotates counterclockwise. Third, you can see how these shedding vortices can suck in air from the vicinity of the wing. Think of the streamlines as lying on a horizontal sheet of paper placed behind the airplane wings, only this sheet is curled up at the left and right edges.

Figure A.11 shows how the pressure differences above and below an airplane wing (which cause lift) give rise to tip vortices that shed from the wings, spilling backwards. My arguments showing the existence of these vortices are intuitive, based upon pressure difference and common sense. Rigorous mathematical derivation of tip vortices is possible, but is root-canal painful. There is a lot of experimental evidence to show that these vortices exist. They contain energy and persist for some considerable time and for some considerable distance behind the airplane, particularly if it is a big one. For this reason, small airplanes must wait a while before following a large plane down onto a runway; otherwise, they might get flipped over. The fact that shedding vortices take energy away from the plane means that the plane experiences a drag—induced drag, first mentioned in endnote 3 of chapter 2.

The same situation applies for boat keels and sails. In chapter 6 I mentioned that some keels (and some airplane wings) are provided with small winglets attached to the tip. These winglets extend in a direction that is perpendicular to the main wing, and they serve to reduce drag and increase lift. In figure A.12 I show you an intuitive argument for why winglets work (and also for why high-aspect-ratio wings work better than low-aspect-ratio wings). Basically, the formation of tip vortices is part of the process of pressure equalization at the wing tip. Because lift force requires a pressure difference, the lift force is reduced near the wing tips. So, as you can see in figure A.12, high-aspect wings work better because the wing-tip pressure reduction is less than it is for low-aspect wings. Adding winglets impedes the wrap-around of air at the tip, and so impedes pressure equalization and lift loss. It also impedes vortex formation and thus limits induced drag.

What about boat sails? The shedding of tip vortices from a fore-and-aft sail is illustrated in figure A.13. Note that, unlike the airplane with its symmetrical wings, the tip vortices are not simply mirror images. The lower vortex is influenced by the presence of the water surface. The overall shape of the tip vortex sheet depends upon sail shape, angle of attack, wind-speed profile, and everything else that you can think of (heeling, foresails, . . .). You will not be surprised to hear me repeat my earlier statement that boat aerodynamics is more complicated than airplane aerodynamics (because of the water boundary and because the sail is not rigid).

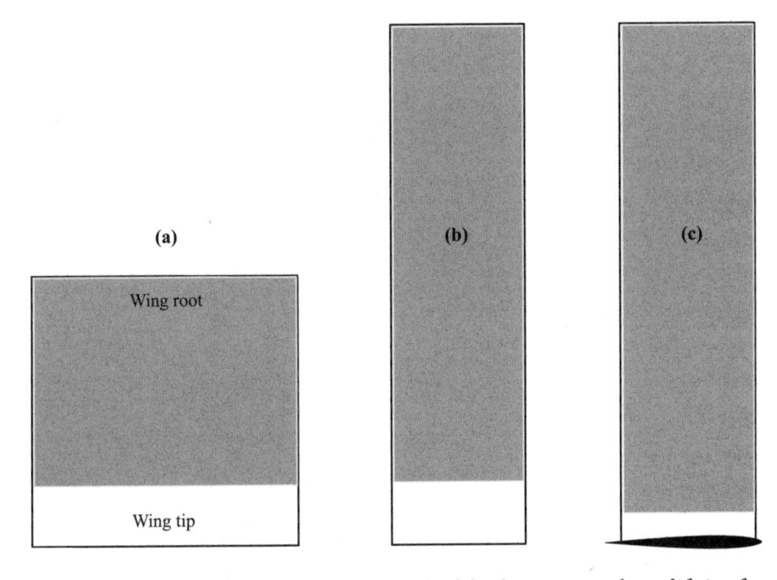

Figure A.12. Rectangular wings or boat keels. (a) The region where lift is effective is shown shaded; the white area near the tip is not effective at providing lift because wingtip vortices here reduce the pressure difference that gives rise to lift. (b) A high-aspect-ratio wing with the same physical area as (a) provides more lift than (a) because less area is affected by wing-tip vortices. (c) Applying a winglet at the tip reduces the flow of fluid from the low-pressure region to the high-pressure region, around the tip, providing more lift than (b) with less drag.

The 3-D shape of sails is described by various parameters, all of which influence the size and shape of the tip vortices. *Camber* is the curved shape of the sail (or wing) airfoil cross section. *Twist* we have already met: it describes the changing (with height up the mast) angle of a sail relative to the centerline. Because the wind speed increases with height, twist is required to maintain a constant angle of attack at different heights. *Taper* refers to the changing shape of a sail with height; the overall shape when seen from the side is called the sail *planform*. All of these factors influence *spanload*, which is the distribution of lift force with distance along the wing or with sail height above the deck. Square sails generate a lot of vortices and thus experience a lot of drag; this is why square sails are good for downwind sailing. Triangular fore-and-aft sails generate smaller tip vortices and so experience less drag. The best planform, for the case of a rigid sail or wing in uniform flow, is elliptical or rather a semi-ellipse like a Spitfire wing. Such a sail has the maximum ratio of lift to induced drag (and induced drag accounts for about three-quarters of the aerodynamic drag acting upon a sail). Look back at some of the high-performance boats pictured in the main text (fig. 4.8, for example), and you will see that their fore-and-aft mainsail looks more

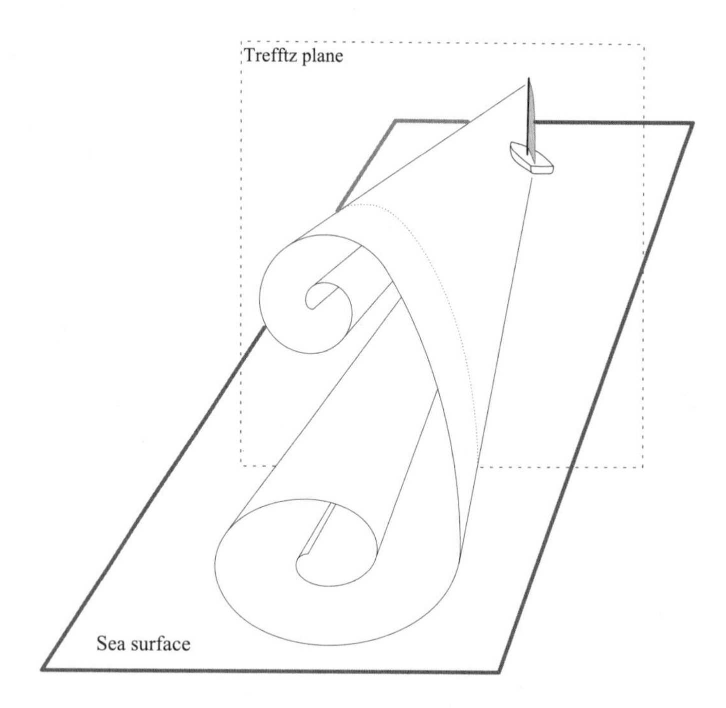

Figure A.13. Tip vortices for a fore-and-aft sail shown way downwind of the boat. The shape of these scrolls depends sensitively upon sailplan, upon the gap between sail foot and boat deck, and upon angle of attack. The sea surface also greatly influences the lower vortex. The *Trefftz plane* is perpendicular to both the sea surface and the apparent wind direction. One is shown by the dashed lines, another is defined by the paper. It is on this imaginary plane that we trace the shape of the vortices.

like a semi-ellipse than a straight-edged triangle. In the real world sails are not rigid, sails must adapt to changing wind conditions such as varying wind speed profile, sails must work in combination with other sails, etc. So, in the real world there is no single sail shape that is best for all circumstances.

Note that I have not explained *why* triangular fore-and-aft sails perform better into the wind than do square-rigged sails.* I have simply stated without proof that fore-and-aft sails (which are basically of triangular planform because they are defined by mast and stays) generate smaller tip vortices than do square-rigged sails. This statement is equivalent to the earlier assertion by the

*You might get an idea from fig. A.12. Because the tip of a triangular wing or sail is a point, less area is affected by pressure equalization, and so lift is not compromised as much as for a rectangular planform. This heuristic explanation is basically right, but the reality is much more complicated.

momentum flux proponents (again made without proof) that the lift and drag coefficients for fore-and-aft sails are different from those of square sails.[2] I have delved deeper only by one layer of explanation: drag is caused by tip vortices, and so reducing these vortices will reduce drag. To understand why triangular sails or elliptical sails generate smaller vortices, I need to resort to mathematics. At this point Professor Feynman would, in a loud stage whisper, tell you that I don't really understand vortex shedding. In his sense this is true, because I can take my intuitive, nonmathematical explanation no further. Those masochists among you who crave for a more detailed explanation, who want to delve into a deeper layer of technical appreciation, are referred to more technical books about sailing fluid dynamics (some of them listed in the bibliography) because in this book I will not take you there. You see how hard it is to explain sailing aerodynamics?

To summarize briefly. Figures A.11–A.13 provide simple physical arguments that air flowing over a wing generates vortices at the wing tips. These tip vortices are shed and carry away energy from the airplane. This loss of energy is experienced as drag. The same idea applies to boat keels and sails. The size of tip vortices, and so the induced drag, depends upon sail planform and other shape parameters.

Flettner and Magnus

I add this section just for completeness. Flettner rotors and the Magnus effect have no direct bearing on sail lift. However, they are a part of fluid dynamics, and a couple of ships have been powered by the force of the wind in this way, so read on. Suppose, instead of considering the circulation of air about the airfoil, we consider airfoils that are circulating in the air. If this rather flippant way of introducing the bizarre Flettner rotor suggests that I am talking about two faces of the same coin, well, I suppose there is a sense in which this may be true. You decide.

The Flettner rotor* was developed in the 1920s as a radically different way to propel boats and ships via wind power. A vessel propelled by this strange mechanism is shown in figure A.14. It looks more like a giant floating candlestick than a boat. The idea is that the vertical column, which has a roughened surface, rotates and carries air with it as a result of the Coanda effect. Here, we have another example of Galilean relativity: it doesn't matter if the air moves and the airfoil surface is stationary, or if the air is stationary and the airfoil moves. All that the Coanda effect requires is relative movement between air and airfoil. In fact, in our frame of reference both air and airfoil are

*Named after Anton Flettner, a German aviation engineer.

Figure A.14. The converted schooner *Buckau* with Flettner rotors attached. This vessel crossed the Atlantic. Image from Wikipedia.

moving: the air moves at the wind speed, and the rotor moves at a controllable rotation rate.

The effect of the rotating cylinder on the airflow is illustrated in figure A.15. The streamlines become curved as the airflow is deflected by the cylinder. Why does rotation cause airflow deflection? There would be no *net* deflection without rotation, after all. The explanation is as follows. The thin boundary layer of air that is carried along the cylinder surface becomes detached when the airflow speed relative to the cylinder becomes too great, just as we expect from the Coanda effect (fig. A.7b). When the cylinder is not rotating, the separation points are symmetric, one on either side of the cylinder. When the cylinder is rotating, however, the separation points migrate to the surface region where relative air speed is highest—in the case of figure A.15, this is the aft portion of the cylinder. In detail, the air separates from the cylinder as vortices, which create circulation as we saw in figure A.4. Calculating the lift that arises from this circulation field yields a force in the direction shown in figure A.15. The ship heading can be adjusted by altering cylinder rotation speed as well as by steering the rudder.

I have given this brief explanation of how a Flettner rotor generates lift force—with the boundary layer peeling off asymmetrically—from the Euler/Bernoulli viewpoint. The same mechanism, given the name of the *Magnus effect,* after the German engineer who proposed it, can be explained via the naïve momentum flux approach as follows: "*Airflow is deflected by the rotating cylinder and so, from Newton's Third Law, the ship experiences a reaction oppos-*

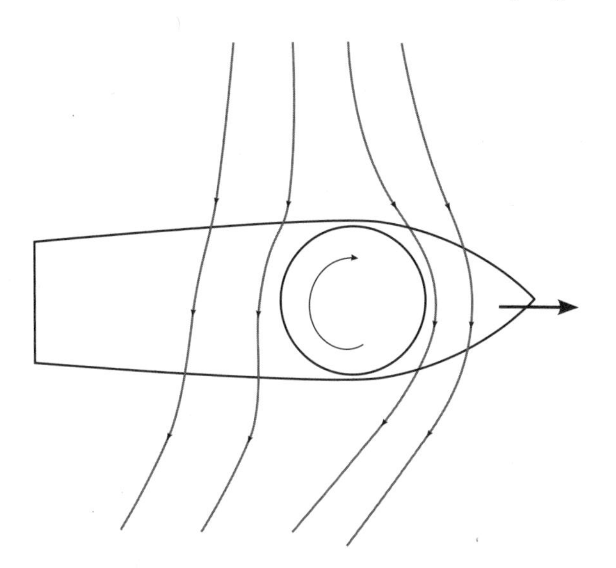

Figure A.15. The Flettner rotor. A rotating drum replaces the sail. The drum drags air around its surface (Coanda effect), but this air detaches where the relative speed of airflow and drum surface is high (here, the aft section of the rotor), creating an asymmetric flow (Magnus effect). From the momentum flux perspective, the airflow is deflected, and lift is provided in the direction shown.

ing this deflection, which is the lift force." The more sophisticated version of momentum flux can fill in some details: *"The rotating cylinder causes streamlines to curve because of the Coanda effect; the curvature gives rise to lift in the observed direction by generating pressure gradients."* Whichever explanation you prefer, the effect is real enough. Flettner was able to propel a ship across the Atlantic in 1926 using two 50-ft-diameter rotors. He found that the vessel worked well to windward and in bad weather. Unfortunately, it was less efficient than conventional steamships or sailing vessels, and so Flettner rotors were never commercially viable.

But the physics is interesting. A number of popular-science articles explain the Magnus effect (which is responsible for the curve of a baseball or soccer ball in flight,* as well as for Flettner rotor motion), and several websites present video demonstrations of it (see the bibliography for details of these).

*"Bend it like Magnus." I doubt if David Beckham, the English soccer star famous for his ability to send a soccer ball on a curved trajectory past helpless goalkeepers, does circulation calculations prior to each shot. He applies "English"—spin—to the ball, and Magnus does the rest.

Notes

1. Evolution: From Prehistory to the Age of Sail

1. A body that is immersed in a fluid, such as water, is buoyed up by a force that equals the weight of displaced fluid. Thus, a beer bottle can float even though glass is denser than water (though a spherical marble made out of the same glass will sink).

2. Let us say that a ship travels a distance, D, of 10 miles upwind: because it is tacking all the way, the distance that it moves over the water, L, is, of course, much greater than 10 miles. With high school trigonometry it is not hard to show that $L = 2D/\sin(2\phi)$, where ϕ is the heading angle into the wind. It is this distance over the water that is reduced by sailing closer to the wind.

2. Analysis: Square-Rigged Ship Motion

1. Physicists know that Einstein's theory of relativity is, in fact, incompatible with Galileo's, though the difference is miniscule at ordinary velocities. Our everyday experience of relativity gives us an intuitive feeling that Galileo was right, but at velocities approaching that of light, things go haywire and Einstein's version takes over. However, unless our sailing boat is skimming across the sea at speeds exceeding a couple of hundred million knots (the speed of light is 583 megaknots), we need not bother ourselves with the discrepancy.

2. According to Sir Isaac Newton, force is rate of change of momentum, where momentum is mass multiplied by velocity. So the magnitude of force is the rate of change of mass multiplied by speed, or $d/dt(mw')$. Here m is the mass of wind that impinges on the boat. If w' is constant, then $F_{wind} = \dot{m}w'$,

where the dot indicates rate of change with respect to time. (This is the original notation used by Sir Isaac when he invented his version of calculus. It is still used by physicists today to indicate time rate of change.) For a sail of area A it is easy to show that $\dot{m} = \rho A w'$, and so $F_{wind} = \rho A w'^2$, as advertised.

3. For a technical explanation, or justification, I turn to Prof. Ludwig Prandtl, the German father of modern aerodynamics. (His revered name will crop up again before the end of this book.) At the beginning of the twentieth century Prandtl and his team developed quite considerably our theoretical understanding of both aerodynamics and hydrodynamics and provided much experimental data with which to test these theories. In his 1934 textbook Prandtl discusses the drag coefficient of various aerofoil shapes. (We may regard a sail as a complicated form of airfoil, the complication arising from the nonrigid structure of sails.) I do not want to get ahead of myself—lift and drag will not be formally introduced in this book until a later chapter. Suffice it to say that long, thin airfoils exhibit less drag than wider airfoils of the same area. The good professor explains that there are three different types of drag: skin friction due to airfoil surface texture, pressure drag due to eddies in the wake behind the wing, and induced drag due to vortices that spill out from the wing tips. Skin friction increases with airfoil roughness and surface area; pressure drag depends on the airfoil profile (i.e., the cross-sectional shape); and induced drag depends on the wing tip details. The last of these is pretty much independent of the wing length, and so is the skin friction for a constant wing area. On the other hand, lift increases with wing length, and so the ratio of lift force to drag force is greater for long, thin wings than for short, thick wings. Thus, wing length (sail height) is the most important dimension for determining lift force.

4. For slow-moving objects, drag may depend on speed in a different way: $F_{drag} = -mbv$. In fact, a detailed analysis shows that the drag constant, b (closely related to the drag coefficient discussed in the appendix), itself depends on speed. However, this dependence is very gradual, and so b can be considered constant over a wide range of speeds.

5. The force equation can be written as a differential equation (the "equation of motion") with boat speed as the variable:

$$\dot{v} = \frac{\rho A}{m} \sqrt{v^2 + w^2 - 2vw\cos(a_{vw})}\,\cos(a_{sw'})[w\cos(a_{vw} - a_{sw'}) - v\cos(a_{sw'})] - bv^2.$$

Here w is true speed, assumed constant. Also, I will assume that the boat moves in a straight line (constant a_{vw}) and maintains a fixed sail trim (constant $a_{sw'}$). The trigonometric factors come from the projections of wind force onto the sail direction and velocity direction, shown in figure 2.3b. Solving the differential equation is a pain, even with simplifying assumptions, but the equilibrium speed (corresponding to $\dot{v} = 0$, i.e. no acceleration) can be found without resorting to calculus.

6. The initial acceleration is found by calculating $(d\dot{v}/dt)|_{v=0}$, with $\dot{v}$ given by the equation of motion in note 4. The condition $(d\dot{v}/dt)|_{v=0} = 0$ separates regions of increasing and decreasing initial acceleration; this condition is found to be $\tan(a_{sw'}) = -[1+\cos^2(a_{vw})]/\sin(a_{vw})\cos(a_{vw})$, from which we obtain the plot of fig. 2.9.

7. The equation of motion in note 5 can be integrated for $a_{vw} = 0$, yielding the solution

$$v = w\,\frac{1 - \exp(-t/\tau)}{\sqrt{\beta} + 1 + (\sqrt{\beta} - 1)\exp(-t/\tau)}$$

which increases from $v = 0$ at time $t = 0$ to $v = v_{eq}$ asymptotically. Here τ is the characteristic time, given as eq. (2) in the main text, which effectively determines how quickly the equilibrium speed is approached.

8. The work done by the wind to propel a boat downwind (from a standing start to speed v) is $W = F_{wind}s$, where s is the distance required to accelerate the boat to speed v. The power imparted by the wind to the boat, $P = \dot{W} = F_{wind}v = \rho A w'^2 v$, increases as the square of apparent wind speed w', as stated in the text.

9. If $a_{sw'} = a_{vw}/2$ then equilibrium speed is

$$v_{eq} \approx \frac{w}{\sqrt{\beta}}\cos(\tfrac{1}{2}a_{vw})$$

and so the upwind speed is

$$v_{eq}\cos(a_{vw}) \approx \frac{w}{\sqrt{\beta}}\left|\cos(\tfrac{1}{2}a_{vw})\cos(a_{vw})\right|,$$

which attains its maximum value for $a_{vw} = 132°$.

10. Had I adopted a different approximation—say the Bernoulli equation—as my guiding principle on sailing ship motion, no doubt I would have obtained a different though quite reasonable approximation to observed ship behavior. Momentum flux is the better introduction, however, because it is simpler to calculate and envisage.

3. Evolution: From the Age of Sail to the Modern Yacht

1. One of the most important military analysts, the engineer and mathematician F. W. Lanchester (1868–1946), proposed his famous (to operational research analysts) "N² Law" at the beginning of the twentieth century. (Lan-

chester was an Englishman but of such international reputation that the Operational Research Society of America still awards an annual prize named after him.) This law states that the fighting strength of a military unit—an army division or a navy fleet—depends on the square of its numbers multiplied by the fighting strength of each member. From this law, Lanchester was able to analyze Nelson's strategy at Trafalgar and found that it was close to optimum. Even 200 years after the battle, Trafalgar tactics are still being analyzed (see Tratteur and Virgilio); modern computer simulations show that Nelson's tactics were also the safest.

4. Analysis: Fore-and-Aft Boat Motion

1. Vectors are underlined; their magnitudes are not. Thus, boat velocity is $\underline{v}$ and boat speed is v. The apparent wind speed is w'. The angle between two vectors is generally denoted by the vector names, so that $a_{vw'}$ is the angle between boat velocity direction and apparent wind direction.

2. The angle $a_{sw'}$ of fig. 2.3 is related to angle of attack, a, as follows: $a_{sw'} + a = 90°$. In terms of a, the sail force of chap. 2 is easily seen to be $F_{sail} = F_{wind} \sin(a)$. If this force is resolved into perpendicular components L and D, then, from fig. 4.3 and eq. (4.1), we obtain eq. (4.2) for lift and drag coefficients.

3. For ease of reading I make an exception here in my notation, by omitting the vector subscripts that define the angle of attack, because I shall be referring to this angle a lot. I note here for later consideration (and without further comment) that, for small angles of attack, it is easy to show that the drag force is proportional to lift force squared.

4. Equation (2.1/4.3) becomes $m\dot{v} = L\sin(a_{vw'}) + D\cos(a_{vw'}) - mbv^2$. As earlier, the last term represents hydrodynamic drag. The first two terms are the contributions of lift and drag to the boat force, acting along the boat velocity direction (see fig. 4.1c). This differential equation can be cast into a form that involves the angle a_{vw} between boat velocity and true wind direction:

$$\frac{\dot{v}}{w} = \tfrac{1}{2}a\left[c_L \sin(a_{vw}) + c_D\left(\cos(a_{vw}) - \frac{v}{w}\right)\right]\sqrt{1 + \left(\frac{v}{w}\right)^2 - 2\,\frac{v}{w}\cos(a_{vw})} - \alpha\beta\left(\frac{v}{w}\right)^2,$$

where w is the constant true wind speed, $\alpha = \rho Aw/m$ and, as earlier, $\beta = mb/\rho A$.

5. Water molecules, and those of most other liquids, will stick to the surface of the hull, no matter how smooth it may be. So as a boat hull moves, it must shear the water molecules—i.e., overcome the forces binding the molecules together (*van der Waals forces*). The energy required to overcome these forces reduces the energy driving the hull forward. We observe this energy-sapping effect as hydrodynamic drag.

6. Recall that equilibrium speed corresponds to speed with zero acceleration, so that $\dot{v} = 0$ in the equation of motion, given in note 4. This reduces the differential equation to an algebraic equation, which can be solved explicitly for certain cases.

7. I.e., by maximizing v_{eq} [eq. (4.4)] with respect to a_{vw}, given constant L/D ratio.

8. You may recall that we found the best sail setting for a square-rigger to be given by $a_{sw'} = a_{vw}/2$. In terms of angle of attack this becomes $a = 90° - a_{vw}/2$, which goes to zero as the boat turns to face the wind ($a_{vw} = 180°$).

9. Because boat speed is small compared with wind speed ($v \ll w$) for realistic values of hydrodynamic drag, I can neglect terms of order v^2/w^2 and so obtain an analytic expression for equilibrium speed. In fig. 4.5 I assume that two different strategies are used (two different "mathematical helmsmen," H1 and H2) for trimming the sail. H1 is used to square-riggers, and he adopts the best sail angle of attack that we determined earlier for square-riggers: $a = 90° - a_{vw}/2$. H2 adopts the following form: $a = 90° - a_{vw}/(1 + a_{vw}/180°)$. Of course, real helmsmen don't sit down with a calculator and set sails according to such trite formulas, but H1 and H2 do serve to show us how sailing performance varies with different choices of sail trim.

10. Known as "jerk" to engineers. So the first three derivatives of displacement are velocity, acceleration, and jerk.

11. This wind-speed distribution with height above the water surface has been parameterized, which is to say it has been measured and then fitted to an equation, $w(h) = w_0 \ln(h/h_0)$. Here h is height above the water surface, h_0 is a length that characterizes the roughness of the water surface, and $w_0 = 0.8\ \mathrm{ms}^{-1}$. The formula holds for light to moderate winds; the logarithm shows that wind-speed gradient lessens with height.

5. A Lot of Torque

1. We construct lines from each vertex to the middle of the opposite side and look for the point of intersection. For two sails, as in fig. 5.2, the CE of each sail is found. The overall CE is somewhere along the line joining them, depending on the relative size of the sails.

2. For such a hull cross section, there is little restoring force (called the *righting moment*) because of the shifting center of buoyancy, discussed later in this chapter. Including hull righting moment would significantly complicate the calculation that I am about to present to you, and I don't want to obscure the woods with the trees.

3. The counterclockwise heeling torque of fig. 5.4 is $\rho A w'^2 h \cos(a_{mz})$, where, you may recall, $\rho A w'^2$ is the force of the wind (assumed to be acting in a hori-

zontal direction). The clockwise torque due to the keel is $m_{keel}gh_{keel}\sin(a_{mz})$, where $m_{keel}g$ is the keel weight. These torques balance at the equilibrium heeling angle, which yields equation (5.1).

4. Those of you who are familiar with engineering or physics will recognize that here I am describing the moment of inertia of rotating objects. The moment of inertia of a hollow cylinder is $I = MR^2$, whereas that of a solid cylinder is $I \equiv MR_g^2 = MR^2/2$. Thus, here the radius of gyration is $R_g = R/\sqrt{2}$.

5. From fig. 5.9 the buoyancy torque about CG leads us to the equation of motion $I\ddot{a}_{mz} = -Mgh_{GM}\sin(a_{mz})$. Here $I = MR_g^2$ is boat moment of inertia, M is boat mass, R_g is radius of gyration, h_{GM} is metacentric height, and g is the acceleration due to gravity. For small roll angles, a_{mz}, this is the equation of simple harmonic motion with angular frequency given by $\omega^2 = Mgh_{GM}/I$, and period (or roll time) $T = 2\pi/\omega = 2\pi R_g/\sqrt{gh_{GM}}$, as in equation (5.2).

6. We model the water wave as a sinusoid: $w(t) = h\sin(\omega t)$, where $\omega = 2\pi c/\lambda$ and wave speed c is given by the deep-water expression $c^2 = g\lambda/2\pi$. Substituting for c gives equation (5.4) for w. From this wave model the angle a_{CB} of fig. 5.11 is easily seen to oscillate as follows: $a_{CB} \approx (2\pi h/\lambda)\cos(\omega t)$. I have made a couple of simplifying approximations here: (1) water waves are not really sinusoidal, but close enough; (2) I assume that the wave slope is small. From fig. 5.11 the righting torque can be written as $MR_g^2\ddot{a}_{mz} \approx Mg.GZ$, where GZ is the righting arm. From fig. 5.11 $GZ = h_{CB}\sin(a_{CB}) - h_{CG}\sin(a_{mz})$. For small angles this expression leads to the equation of motion: $\ddot{a}_{mz} + b\dot{a}_{mz} + \Omega_0^2 a_{mz} \approx \Omega_1^2\cos(\omega t)$, with $\Omega_{0,1}$ given by equation (5.5). As before, I have included a linear (because low-speed) damping term. We recognize the equation of motion as forced-damped simple harmonic motion, the solution of which is well known (see, for example, Kibble and Berkshire) and is presented in the main text. My simple mathematical model makes other assumptions: (1) it applies strictly only for the half-cylinder boat of fig. 5.11; (2) the boat width is much less than the wavelength, $R \ll \lambda$, so that the wave slope is well-defined in the vicinity of the boat hull; (3) the wave slope and boat roll angles are small. Only the last of these assumptions is questionable. The model provides a broad-brush indication of what happens; a more realistic calculation with real hull shapes, real wave profiles, and large angles would be horrendous.

6. Flying through Water

1. This is not quite true. There are *capillary* waves—the tiny waves that you see often superimposed on "regular" water waves—with wavelengths of maybe an inch. These guys are governed by the physics of surface tension, rather than gravity, and the formula for their speed is different. Also, the equation given for

c applies strictly for waves in deep water. We will see in chapter 7 that wave speed in shallow water is different.

2. A vertical arrow in fig. 6.2 corresponds to the buoyancy force of the volume of water at a particular location x, measured from the prow. If, in a length interval dx at x this buoyancy force is dB, then the element of counterclockwise torque provided by dB is $dG = (L/2 - x)dB$. If the wave height above mean is denoted $w_0 \cos(2\pi x/\lambda)$, the total torque is proportional to

$$G = w_0 \int_0^L dx\left(\tfrac{1}{2}L - x\right)\cos\left(\tfrac{2\pi}{\lambda}x\right)$$

Here I have assumed that the hull cross section does not change with length, a sweeping simplification, but in the spirit of this book, it works. The integral is evaluated to yield

$$G = w_0\left(\frac{\sin(z)}{2z} - \frac{1-\cos(z)}{z^2}\right)$$

where $z = 2\pi L\ /\ \lambda$. Now, we are assuming that drag increases as the counterclockwise torque increases, and so will assume that drag depends on z in the same way as does G. This means that drag can be plotted against normalized wavelength, λ/L. This plot is shown in fig. 6.3. Not for the first time in this book I have made sweeping assumptions in order to generate a simple model that captures much of the observed behavior.

3. It is easy to see why drag should increase with torque as follows. Increased "backflipping" torque causes a boat to raise her prow and lower her stern (she is trying to climb her bow wave). This increases the effective area of the boat that is presented to the water passing underneath and so increases drag. My simple analysis does not take into account breaking bow waves; modern naval architects are actively investigating the influence of breaking waves (which are complicated beasts, to a physicist) on ships' drag.

4. This endnote and the next are intended for card-carrying physicists. If you are not one but gamely decide to read on anyway, I will reward you with a nonmath summary that gets across the gist of the analysis. First, I can show why the Kelvin envelope opening angle is 38.94°. In endnote fig. (a) we have a boat that has moved at speed v over a distance OS during time t. So, $OS = vt$. The boat generates waves that spread outwards at the phase velocity v_ϕ, which I will now determine. In two dimensions the wave amplitude can be written $W(\underline{r},t) = W_0 \exp(i(\underline{k} \cdot \underline{r} - \omega t))$, where $\underline{k}$ is wavenumber vector and $\underline{r}$ is location on the water surface. Relative to the boat, this wave looks like $W(\underline{r},t) = W_0 \exp(i(\underline{k} \cdot (\underline{r} - \underline{v}t) - \omega t)) = W_0 \exp(i(\underline{k} \cdot \underline{r} - (\omega - \underline{k} \cdot \underline{v})t))$. So, the wave frequency has been Doppler-shifted due to boat velocity, $\underline{v}$. For a *standing wave* (relative to the

boat!) the frequency is zero (so that *W* has no time-dependence) and so $\omega = kv \cos(b)$, where the angle *b* is defined in endnote fig. (a). The phase velocity is thus $v_\phi = \omega/k = v \cos(b)$. The group velocity is half this value for gravity waves in deep water, as any physics text on hydrodynamics will tell you. So after a time *t*, the waves produced by the boat would spread out a distance $r = 1/2\ vt \cos(b)\ /\ 2$. This is the circle shown in endnote figure (a): a wave generated at $t = 0$ at point *O* radiates out into the circle shown, while the boat progresses to point *S*.

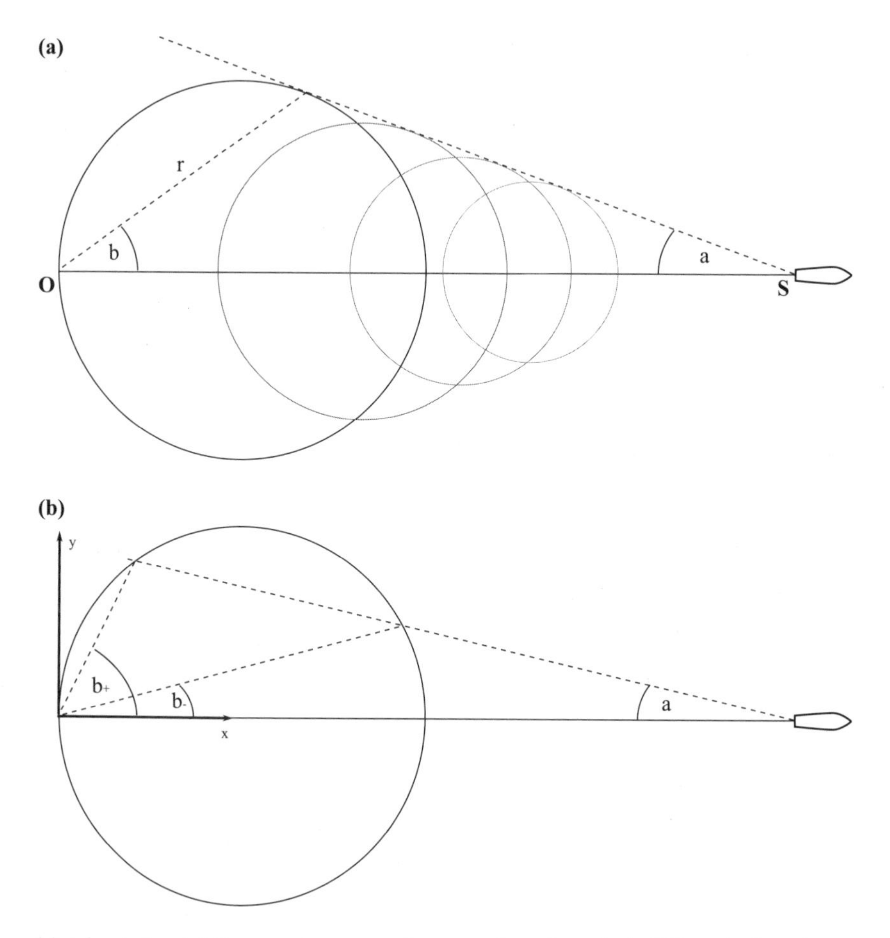

(a) A boat at point O generates waves that spread out in a circle. When the boat has reached point S, the circle is the largest one shown. Waves generated between O and S generate other circles. The envelope of these circles (dashed line) is always at an angle of $a = 19.47°$ to the boat direction. The key feature is that the waves spread at half the boat speed. (b) For an angle, *a*, less than the envelope angle, there are two points on each wave circle that yield stationary waves. These are the points where circle and dashed line intersect.

Because energy travels at the group velocity, these standing waves form a circle with diameter that is only half of the distance *OS*. All along the path of our boat, such waves are generated, as shown. The envelope half-angle is readily seen from the geometry to be given by $\sin(a) = 1/3$ so that the opening angle is $2a = 38.94°$. Summary: Standing waves appear on circles as shown in endnote fig. (a). These waves are left behind the boat rather than keeping up with it because deep water waves travel at half the speed of the boat. The envelope of all the circles generated in this way determines the wake opening angle.

5. Inside the Kelvin envelope, standing waves appear as shown in endnote fig. (b), at the intersection of the circle and the dashed line emanating from the boat. The circle expands along the x and y directions at velocities $\dot{x} = v - v\cos(b)/2$ and $\dot{y} = v\sin(b)/2$. So the distances traveled along x and y directions after time t are $\dot{x}t$ and $\dot{y}t$. From the geometry of endnote fig. (b) we see that the angle a is given by $\tan(a) = \dot{y}t/\dot{x}t = \tan(b/(1 + 2\tan^2(b)))$. Invert this last equation to obtain the two b values in terms of a: $b_{\pm} = (1 \pm \sqrt{1 - 8\tan^2(a)})/4\tan(a)$. These two solutions correspond to the two types of standing wave, transverse and divergent, seen in the wake. There is no solution for $\tan(a) > 1/\sqrt{8}$, so the wake does not extend beyond the Kelvin envelope. (You can check that $2\tan^{-1}(1/\sqrt{8})$ gives us back the opening angle.) Summary: Standing waves inside the Kelvin envelope occur where the circle of endnote fig. (b) and the line at angle a intersect. The evolution of these two points, for all the circles generated as the boat moves through the water, gives us the wave pattern shown in fig. 6.7.

7. Windsurfing

1. The behavior of water waves entering shallower water is analogous to the phenomenon of refraction in optics, when light waves penetrate a medium of increased optical density.

2. Here I might have included the drift velocity, $s\underline{w}$, as part of the sailboard velocity and worked out the apparent wind accordingly. Instead, I choose to regard $s\underline{w}$ as a separate factor to emphasize that it arises from water speed rather than board speed over the water. This preference is a matter of personal taste and is not crucial to the calculation, so long as the drift is taken into account somehow.

3. If the sailboard is drifting downwind (i.e., toward the shore) with velocity $\underline{d}$, then the drift transformation (from yacht velocity to board velocity) is $\underline{v}(\underline{w}) \to \underline{v}(\underline{w} - \underline{d}) + \underline{d}$. Here, $\underline{v}$ is hull or board velocity relative to the water and $\underline{w}$ is wind velocity. The term on the left is yacht velocity. The first term on the right is board velocity relative to the water, whereas both terms on the right represent board velocity relative to land. We are assuming that $\underline{d} = s\underline{w}$ for some

"drift factor" s. The equilibrium speed is calculated from eq (7.1), obtained for ice boats, because we are assuming here (as for ice boats) that hydrodynamic drag is negligible. Thus, the equilibrium yacht speed v_{eq} of eq. (7.1) is the magnitude of $\underline{v}(\underline{w})$. It is straightforward (a geometrical exercise for the interested student, as the textbooks would say) to show that performing the drift transformation changes the magnitude and direction of the sailboard equilibrium velocity as follows: $v_{eq,d} = \sqrt{v_{eq}^2 + 2swv_{eq}\cos(a_{vw}) + s^2w^2}$ and $\tan(a_d) = v_{eq}\sin(a_{vw})/[v_{eq}\cos(a_{vw})+sw]$. Here, $v_{eq,d}$ and a_d are the transformed equilibrium speed and heading angle, respectively. That is, $v_{eq,d}$ is the equilibrium board speed relative to land, and a_d is heading angle relative to land. If there is no drift ($s = 0$), $v_{eq,d}$ reduces to v_{eq}, equilibrium speed relative to the water, and a_d reduces to a_{vw}, heading angle relative to water, as we would expect. In fig. 7.11 $v_{eq,d}(a_d)$ is plotted as bold lines for different s values.

4. We earlier calculated the conditions that must hold for fore-and-aft vessels to create their own wind (fig. 4.6). The same calculation holds for sailboards (it turns out that the drift transformation makes no difference here) except that the lift and drag coefficients vary with angle somewhat differently in the two cases. The range of heading angles for which a sailboard can create its own wind is a little greater than that for a Bermuda-rigged yacht.

Appendix

1. The rate at which fluid mass passes through a Venturi tube is $\dot{m} = \rho vA$, where v is fluid speed and A is tube cross-section area. This rate is constant because the mass leaving the tube per second must equal the mass entering the tube during the same interval. So fluid speeds up when tube diameter decreases (see fig. A.5). For a given streamline, sections with greater speed correspond to less pressure, from Bernoulli's first equation (A.1). Consequently, the pressure is lower in the region of narrower tube. Measuring the pressure difference permits an estimate of the rate at which fluid is entering the tube—hence, the usefulness of the Venturi tube for aircraft.

2. The tip vortex discussion shows how lift is required to generate induced drag (or, if you prefer, both arise from the pressure difference above and below a wing). Recall my earlier statement that, for low angles of attack, drag coefficient is proportional to lift coefficient squared. From this observation you can perhaps begin to see how much of aerodynamics might be explained in terms of vortices, just as much of it can be explained by momentum flux. Aerodynamics is a many-headed beast, like the Hydra of Greek mythology.

Glossary

A comprehensive glossary of sailing terms would double the length of this book. So I will limit entries to technical or historical terms and expressions. Thus, you will not find *sail* here—I assume that anybody interested in this book would already have a pretty good idea what a sail is—but you will find *lateen*. You will not find *cannon,* but you will find *carronade.* In short, I exclude everyday or well-known words. The resulting list is therefore to some degree arbitrary. The aim is to provide a convenient reference for words introduced or widely used in the main text, as an aid to memory.

Age of Sail: The period (late sixteenth to late nineteenth centuries) during which sailing ships dominated international trade and naval warfare.

Airfoil: Cross section of a wing or sail (or propeller).

d'Alembert's paradox: A nonviscous, incompressible flowing fluid exerts zero lift (and drag) upon a body immersed in the fluid.

Angle of attack: The angle presented by a fore-and-aft sail to the wind.

Apparent wind: The wind as felt by a moving ship.

Aspect ratio: The length of a wing or sail divided by its width.

Baltimore clipper: A small schooner or brigantine with raked masts and a hull designed for speed.

Barque/bark: A three-masted sailing ship, with square-rigged foremast and mainmast and fore-and-aft mizzenmast.

Barquentine: A sailing ship with at least three masts, of which only the foremast is square-rigged.

Beam reach: A sailing direction perpendicular to the wind.

Beating: Heading into the wind or beating against the oncoming waves.

Bermuda rig: A vessel with a single mast and with triangular fore-and-aft sail(s).

Bernoulli equation: An equation that expresses the mathematical relationship between fluid flow speed and fluid pressure. Often applied and misapplied in the context of lift and force.

Bilgeboard: A retractable wooden or metal fin (hydrodynamic keel) that is built into each bilge of a ship.

Bireme: An oar-powered warship with two banks of oars.

Board, Chairman of the: The guy with the enormous yacht that inconveniences everybody else in the marina with its large wake.

Boat ownership: "Standing fully-clothed, under a cold shower, tearing up $50 bills." I have been unable to find out who first provided this amusing definition, but it is popular with boat owners and probably goes back several decades. Modern usage often substitutes a larger-denomination bill.

Bonaventure: A fourth mast, aft of the mizzen.

Bowline: A line made fast in the weather leech of square sails to pull them taut and steady when sailing to windward.

Bowsprit: A spar projecting from the bow of a ship to which stays are attached.

Bow wave: A surface water wave generated by a moving ship.

Boyer: A small gaff-rigged merchant ship introduced by the Dutch in the seventeenth century.

Brace angle: The angle that the yards of a square-rigged sailing ship make with the hull. When the yards are perpendicular to the hull's longitudinal axis, the brace angle is zero.

Brig: A two-masted square-rigged sailing ship.

Brigantine: See hermaphrodite brig.

Broach: A sudden change of hull direction upwind due to a rapid increase in sail drive.

Broad reach: A sailing direction that is downwind, but not directly so.

Bulb keel: A hydrodynamic keel with a ballast-filled bulb at the bottom.

Bulkhead: A wall or partition on a boat or ship.

Buss: A shallow-draft Dutch fishing boat with a hinged mast.

Camber: The curvature of a sail, or the asymmetry between the top and bottom curvature of a wing airfoil.

Caravel: A small ocean-going ship of two or three masts used by Portuguese and Spanish explorers of the late fifteenth century. Caravela latina: a lateen-rigged caravel. Caravela redonda: A square-rigged caravel.

Careen: Laying a boat or ship on her side on a beach for below-waterline maintenance.

Carrack: A large three- or four-masted ocean-going sailing ship of the late fifteenth and sixteenth centuries.

Carronade: A short and light smooth-bore cannon of the late eighteenth and early nineteenth centuries developed for use on warships in the Age of Sail.

Carvel: A method of constructing wooden boat and ship hulls, consisting of attaching planks to a frame such that the planks butt up against each other, end to end.

Catamaran: Any boat or ship consisting of two hulls connected by a frame.

Cat boat: A sloop with the mast well forward and usually gaff-rigged.

Centerboard: A movable fin that pivots out of a slot in the center of the hull.

Center of buoyancy (CB): The point within an extended floating body at which the buoyancy force can be considered to act.

Center of effort (CE): The single point at which wind force acting upon a sail or sails can be considered to act.

Center of gravity (CG): The point within an extended body where all the weight or mass can be considered to be concentrated.

Center of lateral resistance (CLR): The point below the waterline at which the hydrodynamic drag force can be considered to act upon the hull of a boat. The geometrical center of the underwater hull profile.

Chordline: A straight line joining the leading and trailing edge of an airfoil, used to define the angle of attack.

Circulation: A measure of vorticity in a flow field. Ignoring viscosity, circulation is proportional to lift.

Clew: The free corner(s) of a sail to which running rigging is attached.

Clinker: A method of constructing wooden boat hulls (in particular) and ship hulls, consisting of attaching planks to each other so that they overlap. Also known as lapstrake.

Clipper: A fast ocean-going merchant ship of the mid-nineteenth century. To some people, the epitome of sailing vessels.

Close haul: A sailing direction, as close to the wind as the vessel can achieve.

Close reach: A sailing direction, a point or two into the wind.

Coanda effect: The tendency of a fluid to follow the curved lines of a nearby solid body, such as the back of a spoon or an airfoil.

Coefficient: Of drag: A dimensionless quantity that characterizes the drag force felt by a body that is submerged in a flowing fluid. Of lift: A dimensionless quantity like the drag coefficient, but pertaining to the lift force. These coefficients depend upon the shape of the body and upon its orientation to the fluid flow.

Cog: A flat-bottomed medieval cargo ship used in northern Europe.

Coming about: See tacking.

Corbita: A sail-powered Roman grain ship.

Courses: The lowest set of sails, nearest the deck, on a square-rigged ship.

Crossing the T: A tactic in the Age of Sail whereby a battle line of ships sailed in front of an enemy line, preferable at right angles and upwind. This tactic permitted more guns to be used in a broadside.

Cutter: A sloop with the mast set well back and with a bowsprit and two or more headsails.

Daggerboard: A long, thin centerboard that slides instead of pivoting.

Difficulty: In this book there are two measures of mathematical impenetrability. Cod-liver-oil: Cod-liver-oil math is so bad that you would rather drink a pint of cod liver oil than work through the math. Root-canal: This is worse: you would rather undergo root-canal treatment than work through the math.

Displacement hull: A hull that moves water aside to make progress. Contrast planing hull.

Doubling: A battle tactic used in the Age of Sail, whereby one line of ships would wrap around an enemy line so that some of the enemy ships received broadsides from both port and starboard.

Downflood angle: The maximum heeling angle, beyond which a boat will flood with water.

Drag: The force that propels a sailing vessel or airplane along the apparent wind direction. More generally, a force generated by passage through a fluid, which resists motion through the fluid.

Drive: The force, derived from the wind, that propels a sailing vessel. Also known as thrust.

Dromon: A Byzantine warship powered by oars and lateen sails.

East Indiaman: A large and slow seventeenth- or eighteenth-century merchant vessel trading between Europe and the Far East.

Effective sail area: The aerodynamic sail area. The effective sail area depends upon sail shape and orientation to airflow as well as upon geometrical area.

Equilibrium speed: The maximum speed that a sailing vessel will reach with a fixed heading and in a wind of constant speed.

Euler equations: The correct, though difficult, way to mathematically describe aerodynamic lift and drag.

Factions: There are two styles of explanation for fluid dynamic lift and drag, referred to in this book as follows: Bernoulli faction: Those who advocate pressure differences and the Bernoulli equation. Momentum flux faction: The other school, who advocate fluid momentum transfer.

Flettner rotor: A strange device for exploiting wind power via rotating air in a vertical, smokestack-like column to drive a ship. Makes use of the Magnus effect.

Flopper-stopper: A deployable float that dampens the rolling motion of a boat.

Fluid: (1) The stuff that boats float on and sail through. (2) The stuff that sailors drink.

Flute: An unarmed Dutch cargo ship of the sixteenth through the eighteenth centuries.

Foot: The lowest edge of a sail.

Fore-and-aft: A sail set more or less parallel to the keel of a boat or ship.

Foremast: The most forward mast of a ship.

Freeboard: The vertical distance from the waterline to the top of the hull (the gunwale) or the deck of a boat.

Frigate: In the Age of Sail, a fast, armed ship that was too small to be a ship of the line.

Fully rigged: Descriptive of a vessel having three or more masts, each square-rigged.

Furl: To roll up and secure; make fast.

Gaff: A spar at an oblique angle from a mast from which is hung a quadrilateral sail aft of the mast.

Galleon: A large multidecked ship (usually a warship) with a high stern, prominent from the sixteenth to the eighteenth centuries.

Galley: An oar-powered vessel used mainly as a warship in the classical world.

Gennaker: A sail that is a cross between an asymmetric spinnaker and a genoa. Gennakers can be used across the wind as well as downwind.

Genoa (jenny): A large jib sail used on Bermuda-rigged vessels for running or reaching in light winds.

Gunwale (pronounced gunnel): The top edge of the sides of a ship's hull.

Halyard: A line to hoist sails or yards.

Head: The toilet, in modern parlance, but also the top of a sail.

Heel: In the nautical context, leaning over at an angle due to wind torque.

Hermaphrodite brig: A two-masted sailing ship with a square-rigged foremast and a fore-and-aft rigged mainmast. Also known as a brigantine.

Hogging: The breaking of a hull on a wave. This occurred in ancient times for long and structurally weak ships when the bow and stern were in the trough of a wave while amidships there was a wave crest.

Hulk: A medieval merchant ship. Later, the term came to apply to a sailing vessel with masts gone, and thus uncontrollable, but still afloat.

Hull speed: The speed of a displacement hull through the water beyond which wave drag increases rapidly.

Hydrofoil: An airfoil in the water. Hydrodynamic keels and rudders are familiar hydrofoils.

Initial stability: The tendency of an upright boat to resist heeling torque. See also range of stability.

In irons: The state in which a sailing vessel is facing directly upwind. While in this state she can make no headway and cannot steer.

Jachtschip ("hunting ship"): A light seventeenth-century Dutch warship. The name is the source of our word *yacht.*

Jibing: A method of sailing upwind; it involves alternating upwind direction by presenting the stern to the wind. This technique is more suitable for square-riggers, and less suitable for fore-and-aft rigs, than tacking. Formerly known as wearing about.

Jib sail: A triangular fore-and-aft sail set forward of the foremast.

Junk: An East-Asian sailing ship, usually associated with China, that was unsurpassed until the nineteenth century by any other type of ship. Characteristics include lugsails and watertight compartments.

Katzmayr effect: The phenomenon that generates thrust from flapping a wing or sail.

Keel: (1) A structure running longitudinally from bow to stern, along the center of the underside, providing the backbone of the frame. (2) A hydrodynamic foil that reduces leeway and adds stability.

Ketch: A two-masted fore-and-aft rigged sailing ship with mizzen mast forward of the rudder post.

Knorr: A northern European clinker-built cargo ship of the late first and early second millennium, double-ended (with a pointed prow at both ends) and with a shallow draft.

Lapstrake: See clinker.

Larboard: An early name for the left side of a ship when facing forward (see also port), referring to the loading ("lading") side.

Lateen: An early triangular fore-and-aft sail, extending forward and aft of the mast, with a long yard at an oblique angle.

Lee: The downwind side of a ship.

Leeboard: A pair of hydrodynamic keels placed symmetrically at the sides of the hull.

Leech: The sides of a square sail; the trailing edge of a fore-and-aft sail.

Lee helm: The tendency of a vessel under sail with a straight rudder to point downwind. See also neutral helm and weather helm.

Leeway: Lateral movement of a boat or ship downwind.

Lift: The component of force perpendicular to fluid velocity direction that is generated by an airfoil or hydrofoil moving through the fluid.

Line: In the nautical context, a rope.

Line of battle: The favored battle formation of Age of Sail warships ("ships of the line").

Long gun: A cannon in the Age of Sail, heavy and long-ranged compared with a carronade.

Longship: A shallow, clinker-built Viking warship with no deck, oar-powered but with a single square sail for open-ocean sailing.

Luff: The leading edge of a fore-and-aft sail.

Luffing: The rippling of a sail (signifying loss of lift) set at too low an angle of attack.

Lug rig: A quadrilateral sail suspended from a yard, like a square-rig, but having an asymmetric yard with more length aft of the mast.

Magnus effect: The shedding of vortices by a rotating body; responsible for curved baseball trajectories and Flettner rotor effectiveness.

Main: (1) The tallest mast, usually near the center of a ship. (2) The Spanish Caribbean.

Metacenter: The intersection of vertical lines through the center of buoyancy of a floating body at equilibrium and heeling.

Metacentric height (GM): A hull characteristic that determines its stability when rolling.

Mizzen: The third mast or the mast immediately aft of the main.

Nao (Nau): Spanish (Portuguese) term for a carrack.

Navier-Stokes equations: The correct but root-canal-difficult mathematical description of fluid dynamics.

Neutral helm: The tendency of a vessel under sail with a straight rudder to stay on course. See also lee helm and weather helm.

Normal: In mathematics, perpendicular.

Ooching: A sudden forward movement by a windsurfer, abruptly stopped, to gain forward momentum.

Pinnace: Historically, a light boat of Dutch origin, a sort of mini-galleon that was used as a tender for larger vessels.

Pitch: (1) A tar-like substance used to waterproof or seal a wooden boat hull. (2) Hull movement (rotation) with the prow moving up and down.

Planform: The shape of a wing as seen from above or below, or of a sail viewed from the side.

Planing: The hydrodynamic lifting of a boat or sailboard hull, significantly reducing the wetted surface and so the hydrodynamic drag.

Planing hull: A hull that moves over the water surface, not through it. Compare displacement hull.

Point: Direction relative to the wind. Also, the compass is divided into 32 points.

Poop deck: The highest deck, at the stern of a ship.

Port: The left side of a ship, when facing forward. Formerly called larboard.

Pumping: The rapid flapping of a sail to increase drive, a common technique of windsurfers.

Quadrireme (quinquireme): An oar-powered warship with four (five) men per oar or per pair of oars.

Quarterdeck: The part of the upper deck of a sailing ship that is aft of the mainmast.

Radius of gyration: For a boat hull, the radius of a cylinder that has the same rotational inertia as the hull.

Range of stability: The maximum heeling angle from which a boat can right herself. See also initial stability.

Reef: (1) The action of reducing a sail's area. (2) A rock or sandbar below the waterline that is hazardous to ships.

Righting arm: The horizontal distance between the center of gravity and the center of buoyancy of a hull. Longer righting arm means greater initial stability.

Righting moment: The torque that acts to restore a boat to its upright position when heeling.

Roll: Hull rotation about the longitudinal axis.

Running: A sailing direction, directly downwind.

Running rigging: Rigging used to move and control sails.

Sailboard: The windsurfer's reason for existing.

Sail plan: A plan drawn to show the dimensions and arrangements of sails to be used in different circumstances for a sailing ship.

Schooner: A fore-and-aft rigged ship usually with two masts, with the tallest second from the front.

Sheet: Part of the running rigging: a line used to control the movable corner(s) of a sail.

Ship of the line: In the Age of Sail, a warship powerful enough to form part of the line of battle.

Shot: Ammunition of an Age-of-Sail cannon. Chain: Two iron balls connected by a chain. Grape: Small iron projectiles loosely attached. Round: Spherical iron ball.

Shroud: Standing rigging that holds up a mast.

Skysail: The topmost sails of a square-rigged ship.

Sloop: A sailing vessel with one mast rigged fore-and-aft.

Slot effect: The improvement in performance of a jib and mainsail combination, over and above the benefits that each provides separately.

Spanload: The distribution of lift force with distance along the wing or with sail height above the deck.

Spar: The general term for a wooden (or metal) sail support. Includes masts, yards, and gaffs.

Spinnaker: A large jib sail (smaller than a genoa) used on Bermuda-rigged vessels for running or reaching in light winds, usually controlled with a spinnaker pole.

Spritsail: A (usually) quadrilateral sail extended by a diagonal spar, which has one end fixed to a mast.

Square-rig: An arrangement in which the main sails are hung from horizontal yards that are more or less perpendicular to the keel.

Standing rigging: Rigging used to support masts and spars.

Standing wave: A wave that does not move; also known as a *stationary wave.*

Starboard: The right side of a ship when facing forward.

Staysail: A fore-and-aft sail set between two masts, or a mast and bowsprit.

Sternpost: The vertical post of a wooden ship, attached to the structural keel, to which the transom (the vertical surface forming the stern) is fixed.

Strake: A large plank used for constructing the wooden hull of a boat or ship, extending from stem to stern.

Streamline: For a flowing fluid, a line that is followed by a particle of the fluid.

Stunsail (from studding sail): A light sail deployed on square-riggers at the ends of a yard, in a light wind.

Surf zone: The near-shore zone where waves build and break.

Swamp: In the nautical context, a hull filling up with water but not sinking.

Tacking: A method of sailing upwind; alternating upwind direction by turning the bow through the wind. More suitable than jibing for fore-and-aft rigs, but unsuitable for square-riggers. Also known as coming about.

Taffrail: The railing around the stern of a ship.

Tail rocker: The upward curve at the back of a sailboard; used to improve maneuverability.

Tall ship: A large, multimasted, traditionally rigged sailing vessel.

Taper: The reduction of wing width (*chord length*) at the tip compared with the wing root.

Teredo navales: Shipworm.

Tern schooner: A schooner with three masts, all fore-and-aft rigged.

Tidal wave: A wave in which the water particles move along with the surface wave.

Topgallant sail: The sail above the topsail but below the skysail on a square-rigger.

Torque: Twisting force. In physics, the product of force and distance from an axis of rotation.

Transient: In physics, an effect that persists only a short time before giving way to a more lasting behavior or condition.

Trapeze: A movement in which sailors hang out of the sides of a boat to counter the heeling torque.

Treenails (pronounced trennels): Long wooden pins employed to connect the planks of a wooden ship's side and bottom to the corresponding timbers.

Trim: (1) Adjustment of sails to the wind. (2) Adjustment of the hull angle (longitudinally) to the waterline by shifting weights within the hull.

Trireme: An oar-powered warship with three banks of oars used by the ancient Greeks.

Tumblehome: The narrowing of a ship's hull above the waterline, seen on galleons and later on ships of the line.

Twist: Changing sail angle of attack with height. A response to changing apparent wind direction with height above the water.

Vang: A brace to steady the mizzen gaff.

Vector: In mathematics, a quantity specified by a magnitude and a direction, represented geometrically by an arrow of specified length.

Velocity: The vector of speed. So, speed plus heading direction.

Venturi nozzle: A constricted tube within which air is accelerated. Used to measure aircraft speed.

Vortex: A spiral flow of fluid.

Vortex shedding: A wingtip phenomenon whereby vortices of fluid break free of the wing surface, taking energy with them.

Wake: The wave disturbance trailing a moving ship.

Wearing: An old-fashioned term for jibing.

Weather helm: The tendency of a vessel under sail with a straight rudder to point upwind. See also lee helm and neutral helm.

Weather side: The windward side of a sailing vessel.

Whipstaff: A vertical stick acting upon the tiller that was part of the steering mechanism of a ship prior to the invention of the ship's wheel.

Wind speed profile: A description of wind speed as a function of altitude above the surface.

Windward: Toward the wind.

Yacht: In modern parlance, almost all vessels sailed for pleasure.

Yard: A horizontal beam supporting a square sail.

Yaw: Hull rotation about a vertical axis.

Yawl: A two-masted fore-and-aft rigged sailing ship with a small mizzen mast aft of the rudder post.

Additional Readings

Sailing History

Anderson, Romola, and R. C. Anderson. *A Short History of the Sailing Ship.* New York: Dover, 2003.

Archibald, Roger. "Six Ships That Shook the World." *Invention and Technology* 13, no. 2 (Fall 1997): 24–37.

A detailed account of the American super-frigates, which include the USS *Constitution.*

Bass, G. *A History of Seafaring.* London: Thames and Hudson, 1972.

Bedini, Silvio A., ed. *Christopher Columbus and the Age of Exploration.* New York: Da Capo Press, 1998.

An inexpensive paperback, this 800-page encyclopedia is a comprehensive reference for the Age of Exploration.

Casson, Lionel. "Rudder, Rigging, Miscellaneous Equipment." Chap. 11 in *Ships and Seamanship in the Ancient World.* Baltimore: Johns Hopkins University Press, 1995.

Cochrane, Thomas, Earl of Dundonald. *The Autobiography of a Seaman: Admiral Lord Cochrane.* London: Chatham, 2000.

A self-serving autobiography by an expert naval commander during the Napoleonic Wars.

Encarta Encyclopedia. Standard Edition. 2005.

Encyclopaedia Britannica. CD 98 Standard Edition. 1998.

Greenhill, Basil. *Archaeology of the Boat.* London: A. & B. Black, 1976. Pp. 139–91.

Hutchinson Encyclopedia. Godalming, UK: Helicon, 1998. S.v. "Ship."

Landels, J. G. "Ships and Sea Transport." Chap. 6 in *Engineering in the Ancient World.* London: Constable, 1978.

Lewis, Jon E., ed. *Life Before the Mast.* New York: Carroll and Graf, 2001.
First-hand accounts of historical battles in the Age of Sail.

Menzies, Gavin. *1421: The Year China Discovered the World.* London: Bantam, 2002.
A controversial and recently debunked, but interesting, view concerning the extent of the giant junks' explorations of the world.

Nicolson, Adam. *Seize the Fire.* New York: Harper Collins, 2005.
Very good historical reference for warships in the Age of Sail, focussed upon the battle of Trafalgar.

O'Brian, Patrick. Historical novels in the Aubrey-Maturin Series.
No reading list about the Age of Sail would be complete without the long series of Aubrey-Maturin novels. O'Brian pays great attention to historical accuracy and nautical technicalities of the period.

Pérez-Mallaína, Pablo E. *Spain's Men of the Sea.* Baltimore: Johns Hopkins University Press, 2005.
Fascinating insight into life onboard the sixteenth-century transatlantic fleet.

Phillips, Carla Rahn. *The Treasure of the* San José. Baltimore: Johns Hopkins University Press, 2007.

Tratteur, Guiseppe, and Raniero Virgilio. "An Agent-Based Computational Model for the Battle of Trafalgar: A Comparison between Analytical and Simulative Methods of Research." In *Twelfth International Conference on Enabling Technologies: Infrastructure for Collaborative Enterprises,* p. 377. IEEE, 2003.

Virtue's Simplified Dictionary: Encyclopedic Edition. London: Virtue, 1948. S.vv. "Rigging and Sails of a Full-Rigged Sailing Ship" and "Types of Sailing Boats and Ships."

Woodman, Richard. *The History of the Ship.* London: Conway Maritime Press, 1997.

Websites

There are a number of interesting websites about individual ships of historical importance. See, for example, the following:

www.hms-victory.com (HMS *Victory* history and pictures)
www.ussconstitution.navy.mil (USS *Constitution* history and pictures)
www.maryrose.org (the English carrack *Mary Rose*)
www.cuttysark.org.uk (on *Cutty Sark,* the last of the clippers)
www.khm.uio.no/english/collections/Viking—ships/oseberg.shtml (the Oseberg longship)

http://sights.seindal.dk/sight/1455—Gokstad—Ship.html (the Gokstad ship and museum)

http://hem.bredband.net/johava/WASA2e.htm (the giant Swedish galleon *Vasa*).

Plymouth City Council. CyberLibrary, section on Naval History: www.plymouth.gov.uk/homepage/leisureandtourism/libraries/cyberlibrary/clbrowsethecyberlibrary/clhistory/clnavalhistory.htm

Sailing Theory and Practice

Anderson, Bryon D. *The Physics of Sailing Explained.* New York: Sheridan House, 2003.

A short technical book, good on induced drag.

Chase, Carl. *Introduction to Nautical Science.* New York: W. W. Norton, 1991.

Practical advice, plus an excellent nonmathematical description of torque effects that influence boat handling.

Garrett, Ross. *The Symmetry of Sailing.* New York: Sheridan House, 1996.

Gentry, Arvel. "A Review of Modern Sail Theory." Proceedings of the Eleventh AIAA Symposium on the Aero/Hydronautics of Sailing, September 12, 1981, Seattle, Washington.

This is a good nonmathematical explanation of sailing aerodynamics and is available online, along with the other references by Gentry, a professional aerodynamicist, at www.arvelgentry.com/.

——. "Boundary Layer Flow and the Headsail." *Sail Magazine,* May 1973.

——. "How a Sail Gives Lift." *Sail Magazine,* June 1973.

——. "Another Look at the Slot Effect." *Sail Magazine,* July 1973.

——. "How Sails Really Work." *Sail Magazine,* April 1983.

Gerr, Dave. *The Nature of Boats.* Camden, Maine: International Marine, 1992.

Much practical advice about sailing and boatbuilding.

Hedges, Martin. *The World of Sailing.* New York: Elsevier-Dutton, 1981.

Killing, Steve, and Doug Hunter. *Yacht Design Explained: A Sailor's Guide to the Principles and Practice of Design.* New York: W. W. Norton, 1998.

Larsson, Lars, and Rolf E. Eliasson. *Principles of Yacht Design.* Camden, Maine: International Marine, 2000.

Detailed technical design principles built up from solid hydrodynamic and aerodynamic foundations.

Marchaj, C. A. *Aero-Hydrodynamics of Sailing.* Camden, Maine: International Marine, 1989.

A modern classic by a prolific and authoritative author in this field. This book contains every technical detail you may care to know about sailing fluid dynamics.

——. *Sail Performance.* Camden, Maine: International Marine, 2003.

A technical but nonmathematical exposition of all aspects of sailing, by an expert with 40 years' research experience.

Whidden, Tom, and Michael Levitt. *The Art and Science of Sails.* New York: St. Martins Press, 1990.

A well-received practical guide to sailboat construction and handling.

Website

Wee, Denis. "The Wind and Apparent Wind," at http://web.singnet.com.sg/~dgswee/AppWind/trueapp.htm.

A good elementary account of the wind forces involved in windsurfing.

Theory Relevant to Sailing

Auerbach, David. "Why Aircraft Fly." *European Journal of Physics* 21 (2000): 289–96.

Babinsky, Holger. "How Do Wings Work?" *Physics Education* 38 (2003): 497–503.

Coulson, C. A. "Waves in Liquids." Chap. 5 in *Waves.* London: Oliver and Boyd, 1958.

Douglas, J. F., and R. D. Matthews. "Buoyancy and Stability of Floating Bodies." Chap. 3 in *Fluid Mechanics.* Harlow, Essex, UK: Longman, 1996.

Basic fluid mechanics for the technically minded, with many calculated examples.

Kibble, T. W. B., and F. H. Berkshire. "Linear Motion." Chap. 2 in *Classical Mechanics.* Harlow, Essex, UK: Longman, 1996.

A standard university undergraduate physics text, this book has a clear presentation of the theory (forced, damped simple harmonic motion), applied here to describe rolling motion.

Massey, B. S. *Mechanics of Fluids.* London: Chapman and Hall, 1997.

A comprehensive technical development of fluid mechanics principles.

Nathan, Alan. "Baseball Pitches." *Scientific American,* September 1997, pp. 83–84.

A well-known application of the Magnus effect.

Prandtl, Ludwig, and O. G. Tietjens. *Applied Hydro- and Aeromechanics.* New York: Dover, 1934.

A classic text, still readable. Prandtl performed much of the early work (both theoretical and experimental) that led to our current understanding of aerodynamics.

Ruggles, K. W. "The Vertical Mean Wind Profile over the Ocean for Light to Moderate Winds." *Journal of Applied Meteorology* 9 (1970): 389–95.

Swartz, Clifford. "Force and Pressure." Chap. 1 in *Back-of-the-Envelope Physics.* Baltimore: Johns Hopkins University Press, 2003.

How to apply the Newtonian momentum transfer model of airfoil lift.

Walker, Jearl. "The Feathery Wake of a Moving Boat Is a Complex Interference Pattern." *Scientific American* 258 (1988): 124–27.

Waltham, Chris. "Flight without Bernoulli." *The Physics Teacher* 36 (1998): 457.

Websites

National Aeronautics and Space Administration, Glenn Research Center website, at www.grc.nasa.gov/WWW/K-12/airplane/wrong1.html.

NASA explains lift, beginning by debunking wrong theories here.

University of Maryland Physics Lecture-Demonstration Facility website, at www.physics.umd.edu/lecdem/services/demos/demosf5/f5–31.htm.

A website with a video clip demonstrating the Magnus effect and the Flettner rotor.

Weltner, Klaus, and Martin Ingelman-Sundberg. "Physics of Flight—Reviewed," at http://user.uni-frankfurt.de/~weltner/Flight/PHYSIC4.htm.

One of a series of educational web articles by aerodynamicists from Frankfurt University.

Index